ACCIDENTAL DISCOVERIES

From Laughing Gas to Dynamite
- Revised & Expanded Edition

LARRY VERSTRAETE

 FriesenPress

Suite 300 - 990 Fort St
Victoria, BC, v8v 3K2
Canada

www.friesenpress.com

Copyright © 2016 by Larry Verstraete
Revised Edition — 2016

Revised and Expanded Edition Previously published by Scholastic Canada Ltd.under the titles: *Accidental Discoveries: From Laughing Gas to Dynamite & The Serendipity Effect*
Copyright 1989, 1999, 2015 by Larry Verstraete

All rights reserved.

No part of this publication may be reproduced in any form, or by any means, electronic or mechanical, including photocopying, recording, or any information browsing, storage, or retrieval system, without permission in writing from FriesenPress.

ISBN
978-1-4602-7721-8 (Paperback)
978-1-4602-7722-5 (eBook)

1. SCIENCE, EXPERIMENTS & PROJECTS

Distributed to the trade by The Ingram Book Company

Dedicated, as always, to my family - with love

And to the memory of my parents, George and Paula

TABLE OF CONTENTS

vii   **Other Books by Larry Verstraete**

ix   **Author's Note**

xi   **Introduction**

1   **Chapter 1**
Aha Moments

27   **Chapter 2**
Fortunate Fumbles

51   **Chapter 3**
Opportunity Knocks

73   **Chapter 4**
Experimental Twists

99   **Chapter 5**
Clever Connections

123   **Chapter 6**
Surprise Endings

143   **Glossary**

147   **For Further Reading**

OTHER BOOKS BY LARRY VERSTRAETE

"Dinosaurs" of the Deep: Discover Prehistoric Marine Life, *Turnstone Press*

Missing in Paradise, *Rebelight Publishing Inc.*

Life or Death: Surviving the Impossible, *Scholastic Canada*

Surviving the Hindenburg, *Sleeping Bear Press*

Case Files: 40 Murders and Mysteries Solved by Science, *Scholastic Canada*

S is for Scientists: A Discovery Alphabet, *Sleeping Bear Press*

At the Edge: Daring Acts in Desperate Times, *Scholastic Canada*

G is for Golden Boy: A Manitoba Alphabet, *Sleeping Bear Press*

Lost Treasures: 25 True Stories of Discovery, *Scholastic Canada*

Survivors: True Death-Defying Escapes, *Scholastic Canada*

Extreme Science: Science in the Danger Zone *Scholastic Canada*

Whose Bright Idea Was It?: True Stories of Invention, *Scholastic Canada*

Mysteries of Time, *Scholastic Canada*

AUTHOR'S NOTE

The concept for this book, like much of its content, stemmed from unexpected circumstances. Years ago, as a newbie writer learning the trade, I signed up for a correspondence writing course in children's literature. The first few assignments involved writing fiction, but the fourth required something different. "Write a non-fiction article for a children's magazine," the instructions said. With a background in science and a teaching career already in full swing, I dipped into a comfortable subject and selected 'lightning', a topic I felt would captivate young readers.

In my research for the article, the name Benjamin Franklin surfaced. So did the famous and familiar story of his dangerous kite-in-a-lightning-storm experiment. Then, as I dug deeper into my research material, I encountered another story about Franklin, one that occurred two years before the kite experiment. It involved a holiday party, a turkey destined for electrocution, a colossal accident on Franklin's part, and a discovery that altered the course of history. Right then, I realized that I'd discovered writer's gold – a story so odd and fascinating that, properly told, it practically guaranteed the reader's attention. I abandoned my earlier subject and wrote about Franklin and the turkey instead.

With my science background, I knew there were other science stories with similar mixes where mishaps, mistakes, and unusual circumstances ultimately led to major breakthroughs. As I worked on other course assignments, I wrote about these, too. By the end of the course, I had a sizeable collection - enough for a decent book.

In 1989, Scholastic Canada published the manuscript under the title *The Serendipity Effect.* Several years later, it was revised and reissued under a new title: *Accidental Discoveries: From Laughing Gas to Dynamite.*

Since that time, the Internet has broadened the scope and accuracy of research, and fortunately modern science still benefits from fruitful blunders and twists of fate. This expanded and updated edition of *Accidental Discoveries* contains more than 80 stories. Many are new. Others like Benjamin Franklin's turkey are timeless favorites. Together, the stories show that in the hands of someone insightful and curious even minor disasters can have silver linings.

Enjoy!

INTRODUCTION

What triggers the brainstorm that leads to a scientific discovery? What stirs an inventor to create something new or to see possibilities never seen before?

Ideas surface in the strangest ways, often when they are least expected. Sometimes they seem to pop up almost by accident.

The story of an unusual discovery in a Kodak research laboratory is a good example.

In 1951, a group of Kodak chemists tried to find a tough, clear, heat-resistant plastic to use in jet plane canopies. One of their tests involved measuring how far light bent as it passed through the plastic. To make the measurement, they used an expensive machine called a refractometer.

Usually the refractometer test was simple and quick. The chemist placed a sample of plastic between two prisms in the machine, shot a beam of light through the plastic, measured how far the beam bent, removed the plastic, and went on to do the next test. One day, however, things did not go as planned. The plastic stuck to the prisms, wedging them in the machine. The refractometer was ruined. The chemist sadly reported the loss to his supervisor, Harry Coover.

At first, Coover was discouraged. No amount of tugging or prying could separate the prisms. The plastic film had fused them together. Then suddenly Coover realized that the loss of the refractometer was not really such a serious loss after all. Although the chemists had been searching for a material to use in jet planes, they had accidentally discovered something almost as valuable – a substance that bonded materials so well that they could not be separated. The discovery led to the development of new types of fast-acting, powerful adhesives called super glues.

Of course, serious scientists and inventors don't depend on accidents for success. But mistakes, mishaps, unusual coincidences, and strange twists of luck happen all the time. Occasionally such surprises can be helpful. Sometimes they provide new and valuable information, point out solutions to problems or open the doors of imagination, making the impossible suddenly seem possible.

Fate has often played a part in science and invention. In fact, we even have a word to describe it. The ability to make unexpected discoveries by accident is called *serendipity*.

This book is about the errors, accidents, coincidences, and odd circumstances that have started or changed the discovery process. It is about creative thinking and what it takes to generate ideas. Above all, it's about inventions and breakthroughs, old and new, large and small, that are due in some way to the *serendipity effect*.

CHAPTER 1

AHA MOMENTS

Have any of these ever happened to you?

- You find a chunk of food that's been lying around too long. It's covered with disgusting mold.

- You fill a basin with water. Then you put something in it and water sloshes all over the floor.

- After a hike in the woods, you discover burrs stuck to your clothes.

- A bookmark falls out of a book you are reading, and you lose your place.

It's likely you've had at least one of these experiences. It's probably just as likely that you've never given it a second thought.
Perhaps you should . . .

In this chapter, each of these perfectly ordinary events triggered a brainstorm and led to a revolutionary discovery or new invention. In the hands of someone observant, to someone curious with lots of questions, everyday experiences like these can become 'aha' moments when flashes of insight seem to appear out of nowhere.

Pythagoras – About 540 B.C.
A BLACKSMITH'S POUNDING

The day was likely warm. Shops were probably busy and the dusty streets swarmed with sandal-clad people. Without precise records from the time, all we can do is guess, but 2500 years ago this was a common scene in places like Creton, a city in southern Italy.

That particular day, one of the people wandering the streets of Creton was a Greek mathematician named Pythagoras. Out for a stroll, Pythagoras was not in a rush. When he passed a blacksmith making horseshoes in the doorway of his shop, he stopped to watch.

The blacksmith stoked a fire and pulled out a red-hot piece of iron. Using a hammer, he pounded and shaped it on an anvil. Each time the blacksmith brought his hammer down, a loud clang filled the air.

At first, the sounds seemed ordinary enough. But then Pythagoras's keen senses noticed something else. Whenever the blacksmith switched anvils, the sounds changed. The tones were different. With his curiosity peaked, Pythagoras pondered the situation as he continued on his way.

At home, he stretched a piece of string between two wooden pegs on a board. He plucked the string and heard a musical twang. When he used a longer string, he heard a lower, deeper twang.

By changing the lengths of strings, Pythagoras made an interesting discovery. The longer the string, the lower the tone. The shorter the string, the higher the tone.

Even more curious now, Pythagoras began a series of investigations. He chose a string and tied it tight to the board. Next to it he tied another string twice its length. When he plucked both strings together, Pythagoras found that they produced a pleasing combination of notes. Because one string was exactly twice as long as the other, their mathematical ratio was 2 to 1.

When Pythagoras used a string that was 1½ times as long as the first, he produced another pleasing combination of notes. This time, the ratio of lengths was 3 to 2.

Over and over, Pythagoras changed the lengths of strings and compared the musical notes he made. He found that the most harmonious or pleasant sounds were made when the lengths were in small ratios to each other – 2 to 1, 3 to 2, 4 to 3. When he tried more complicated ratios – 19 to 9, or 23 to 13 – the sound combinations were less pleasant.

Pythagoras discovered that there is a predictable numerical pattern to the most pleasing musical sounds. By applying their mathematical ratios, he could compose a whole range of harmonious notes.

Since then, the connections between music and mathematics have been studied further, but the numerical values Pythagoras discovered still apply. Nowadays all musical instruments – string, wind, and bass – rely on simple musical ratios like those discovered by Pythagoras many centuries ago.

Archimedes – About 250 B.C.
AN OVERFILLED BATH

Over two thousand years ago, the most feared force on earth was the powerful Roman army. Yet when the mighty Roman troops set out to crush the Greek city of Syracuse, they were almost flattened by huge catapults heaving enormous boulders, mechanical cranes that seized and overturned entire ships, and massive lenses that focused the sun's rays on enemy vessels, setting them on fire.

For their time, these war-machines were truly awesome, but to Archimedes, the inventor of these weapons, they were mere toys, objects for his amusement. Today Archimedes is remembered more for a scientific discovery and the strange story behind it than for his clever machines.

By all accounts, Archimedes was a deep thinker. Often hours flew by as he considered a problem. Then he would suddenly announce a solution as though the answer had just popped into his head. Perhaps the famous story of the king's crown is the best example.

King Hieron II of Syracuse ordered a new crown made out of solid gold. The finished article was beautiful, but Hieron was suspicious. Had the goldsmith mixed silver with the gold and lowered the crown's value by changing its purity? Hieron asked Archimedes to find out the truth without damaging the crown.

Archimedes pondered the problem. Silver is less dense or compact than gold and therefore weighs less. That much he knew. The obvious thing to do was to weigh the crown and then weigh an equal amount of pure gold to see if their weights were the same.

But how could he measure the precise amount of metal that had gone into the making of the crown? The only sure method was to melt down the crown and then measure the volume of the molten liquid. Doing that would destroy the crown, though, and Archimedes was under strict orders not to damage it.

Gradually, Archimedes became more and more possessed by the problem. He lost track of time, forgetting even to eat or sleep. Then one day he went to the public baths to relax. As he stepped into the full bath, he noticed that the water rose.

To everyone else, this was just something that happened every day. But to Archimedes, it was the solution to his problem. In that instant, he realized that the amount of water raised or displaced equaled the volume of his body as he got into the bath.

Archimedes leaped out of the bath so excited he didn't even dress. He raced out of the building and down the street shouting, "Eureka! Eureka!" (I have found it!)

At home, Archimedes pushed the king's crown into a bowl filled to the top with water. He measured the amount of liquid that spilled over. From his experience at the public baths, he knew now that the volume of the water displaced by the crown would be the same as the volume of the crown itself.

Accidental Discoveries 7

Next Archimedes measured out an equal volume of pure gold. Then he checked the weight of the pure gold against the weight of the crown. Sure enough, the crown was lighter. The king had indeed been cheated.

The goldsmith was punished, Archimedes was rewarded, and the world was given a way to establish the relative density and purity of different materials.

DID YOU KNOW?

ARCHIMEDES DIED MUCH THE SAME WAY AS HE LIVED – DEEP IN THOUGHT. WHEN SYRACUSE WAS FINALLY CONQUERED BY ROMAN TROOPS, A SOLDIER WAS SENT TO ARREST HIM. THE SOLDIER FOUND THE OLD MAN SITTING ON THE FLOOR, SOLVING MATHEMATICAL PROBLEMS. WHEN ASKED TO SURRENDER, ARCHIMEDES REFUSED, SAYING THAT HE WAS TOO BUSY WITH HIS WORK. THE ANGRY SOLDIER DREW HIS SWORD AND KILLED ARCHIMEDES ON THE SPOT.

Galileo Galilei – 1581

THE SWINGING CHANDELIER

It was a typical Sunday in 1581. Hundreds of worshippers filled the huge cathedral in Pisa, Italy. Most of them listened intently to the church service.

But not seventeen year-old Galileo Galilei. Instead, Galileo studied a chandelier hanging overhead. Air currents flowing through the lofty cathedral moved the chandelier from side to side, back and forth. Sometimes the chandelier moved gently; sometimes it swung in a wide arc. No matter what the size of its swing, it seemed to Galileo that the chandelier kept steady time.

There were no clocks or watches in those days. To time the chandelier's swings, Galileo felt for the pulse in his wrist. He counted the pulse beats. One, two, three beats for one swing. One, two, three beats for another.

Galileo was surprised. No matter how wide or narrow the swing, it always took the same number of pulse beats.

Right after the service, Galileo raced home. He quickly suspended a weight from a long string to create a pendulum. Galileo pulled the weight back a short distance, released it, and timed its swing. He tried it again, this time pulling the weight back farther before releasing it. After many tries, Galileo confirmed his suspicions – the time it took to make one swing was always the same whether the swing was wide or narrow.

Excited now, Galileo tried other experiments with his pendulum. He discovered that the length of string, amount of weight, and other factors all had some predictable relationship to the time of a pendulum's swing.

Some years later, Galileo experimented with falling objects. Did all objects fall at the same rate? To find out, he needed to time objects as they fell. But that posed a problem. How could he accurately time something that moved so quickly?

Galileo remembered the pendulum. The weight of the pendulum acted just like a falling object – except it didn't fall straight down. It fell on a slant and at a slower rate that could be timed.

Galileo adapted the pendulum as a timepiece. First he got a wooden board and carved a long, straight, smooth groove down the center. When he raised the board slightly at one end and released a ball, it slowly rolled down the groove.

Galileo marked off his grooved board into small divisions of equal length. For a timing device, he rigged up a water-filled container with a small hole in the bottom. By counting water drops, he could keep track of time. Now he was ready to begin.

He released one ball at a time from the higher end of the board. As the balls rolled, Galileo timed how long it took them to cross each division of the board. To his surprise, Galileo discovered that the balls didn't travel down the track at an even rate. Instead, they accelerated – or sped up – as they got farther down the groove. Falling objects, he found, picked up speed as they fell to the earth.

After more experiments, Galileo was able to work out a mathematical formula to calculate the acceleration of a falling object. To prove his point, he even predicted how far a cannonball could be blasted from a cannon. Then he fired it to verify his prediction.

In many ways, the swinging chandelier started a revolution in the world of science. With his pendulum investigations, Galileo pioneered the scientific method —the system of carefully controlled experiments and observations that modern scientists use today to prove a natural law beyond a shadow of a doubt.

Robert Koch – 1880
A MOLDY POTATO

Today we know that microorganisms such as bacteria can cause disease. By controlling the spread of these microorganisms, we can protect ourselves from illness. In the late 1800s, though, this was a brand new idea and many people didn't believe it. How could something too tiny to be seen by the unaided eye actually cause disease?

Robert Koch, a German doctor, did not believe that the idea was ridiculous at all. He spent long hours in his laboratory in Berlin trying to isolate and study bacteria.

One day in 1880, while he was cleaning up his laboratory, Koch noticed a piece of boiled potato someone had left on the table. The potato had been lying there a few days and already it was covered with furry mold.

Koch picked up the potato. As he was about to discard it, he stopped and looked more closely. Wasn't this interesting?

Although Robert Koch had seen moldy food before, this time he noticed something different. Separate patches of mold covered the potato. Each patch was a different color.

Koch pulled off a bit of gray mold, put it on a glass slide, added a drop of water, and looked at it under his microscope. A swarm of identical-looking microorganisms swam across the slide. Next he examined a red patch. Interesting! A different kind of microorganism this time.

Test after test, Koch observed the same thing. Each colored patch contained clusters of identical-looking microorganisms that were different from the clusters of microorganisms growing in other colored patches. When Koch plucked microorganisms from one patch and placed them on a food source, they grew and multiplied, forming a colony of microorganisms identical in every way to the original.

Robert Koch's second glance at a simple potato gave scientists a way to grow and isolate bacteria as well as other microorganisms for study. It was the beginning of the science of bacteriology and the end to many diseases.

> **MORE ACCIDENTAL DISCOVERIES**
>
> WHILE RESEARCHING A NEW ANTI-ULCER DRUG IN 1965, JAMES SCHLATTER, A CHEMIST, SPILLED SOME POWDER ONTO HIS FINGERS. WHEN HE LICKED HIS FINGERS TO PICK UP A SHEET OF PAPER, HE NOTICED A STRONG SWEET TASTE. THE ACCIDENT LED TO THE DISCOVERY OF ASPARTAME, AN ARTIFICIAL SWEETENER.

Frank Epperson – 1905

ACCIDENTALLY FROZEN

The first Popsicle went on sale in 1923. But the real story of the Popsicle started eighteen years earlier with a small boy, a jar of soda water, a stick, and an unusually cold night.

One day in 1905, eleven year-old Frank Epperson of California whipped up a popular drink of his time. He added powdered soda mix to water. By mistake, he left the mixture on his back porch overnight. That night temperatures dropped to an all-time low. The next day Frank discovered that the jar of soda water had frozen with the stirring stick stuck inside.

Years passed. Frank Epperson forgot about the event. As an adult, he tried his hand at several businesses, but none of them was a resounding success. Then, eighteen years after the actual incident, Epperson remembered the frozen soda water of his youth. He changed the recipe by adding fruit flavors in place of the soda water, then poured the brew into a mold, added a handy wooden carrying stick, and froze the mixture. Epperson called the first of these frozen treats "Epsicles" for Epperson's Icicles. Later the name was changed to Popsicle.

Today two billion Popsicles in over twenty-six flavors are sold each year. The favorite? The best selling flavor is cherry.

DID YOU KNOW?

THE NAME "POPSICLE" ACTUALLY CAME FROM FRANK EPPERSON'S OWN CHILDREN. THEY NICKNAMED THE ICE TREAT "POP'S CYCLES". WHEN THE GREAT DEPRESSION HIT NORTH AMERICA IN THE 1920S, PEOPLE COULDN'T AFFORD TREATS LIKE POPSICLES. TO COMBAT SLIPPING SALES, EPPERSON INTRODUCED THE TWIN POPSICLE – TWO POPSICLES FROZEN TOGETHER IN ONE PACKAGE. FOR A NICKEL, A KID COULD BUY A DOUBLE-SIZED TREAT AND SHARE IT WITH A FRIEND. ORIGINALLY EPPERSON FROZE EACH POPSICLE SEPARATELY IN LARGE TEST TUBES, A PROCESS WHICH TOOK SEVERAL HOURS. WITH MODERN EQUIPMENT, IT NOW TAKES MINUTES.

MORE ACCIDENTAL DISCOVERIES

More than one sugary treat owes its shape and name to an accidental discovery:

LIFESAVERS – The candy's distinctive hole-in-the-middle shape came about when Clarence A. Crane, its creator, hired someone to mass-produce a flat, round candy. Instead, the machine malfunctioned and punched a hole in the center. Crane liked the unexpected effect and gave the candy its well-known name.

MILK DUDS – In 1926, when Chicago candy manufacturer E. Hoffman & Company tried to develop a perfectly round, chocolate-covered caramel, the machine spit out lopsided balls instead. The mangled milk chocolate 'duds' tasted so great that the company continued production. The deformed candy was marketed under the name Milk Duds.

Sylvan Goldman – 1936

CHAIRS + WHEELS = ?

If you went grocery shopping before 1937, you needed a sturdy back and strong arm muscles. As you roamed the aisles, you piled groceries into baskets and bags that you lugged around the store with you. Strenuous exercise!

Sylvan N. Goldman of Oklahoma City was in the supermarket business. He figured that there had to be an easier way.

One night in 1936, Goldman sat on a folding chair in his small office thinking about this problem. How could he make shopping more convenient and less exhausting? He toyed with several plans, but none of them was very good. Then, as he glanced at his chair, he had a whopper of an idea. Later he said, "Inspiration hit me right between the eyes."

He placed two folding chairs together and studied them. If he joined the chairs and added wheels underneath and a basket on top, Goldman figured that he would have an easy-to-push cart. Like the chairs, the cart could be folded for convenient storage, too. Goldman called his new invention the "folding basket carrier".

On June 4, 1937, Goldman's first batch of shopping carts stood ready for use in his store. But the day was a disappointment. Instead of excitedly grabbing carts, customers avoided the new invention believing that other shoppers would think they were weak if they used one.

Then Goldman had another great idea. He hired people to push carts around this store, pretending to be shoppers. It wasn't long before real customers copied the phony ones and followed their example.

Today hundreds of millions of shopping carts roll around stores taking the strain out of lugging supplies for customers around the globe.

> **MORE ACCIDENTAL DISCOVERIES**
>
> CHEDDAR, ANYONE? ACCORDING TO LEGEND, A WANDERING ARAB TRAVELING THROUGH THE DESERT OPENED A POUCH CONTAINING MILK, ONLY TO FIND THAT THE LIQUID HAD SPOILED IN THE HEAT AND HAD SEPARATED INTO THICK CLUMPS. THROUGH A TWIST OF FATE, HE HAD DISCOVERED A WAY TO TURN MILK INTO CHEESE.

James A. Crocker – 1990
SAVING HUBBLE

In the history of colossal mistakes perhaps none was more costly or embarrassing than the Hubble Space Telescope. Hubble was launched into space in April 1990, after forty-four years of planning and development that cost a cool $1.5 billion. As Hubble orbited Earth, its polished lenses and mirrors were supposed to transmit crisp images of distant stars, allowing astronomers to peer into far-away galaxies. Instead, the images transmitted by Hubble were fuzzy and disappointing.

Red-faced engineers at NASA quickly diagnosed the cause. Hubble's 2.4 meter (8 ft.) primary mirror had been polished into the wrong shape. Its outer edges were off roughly 1/50th the thickness of a human hair – just enough to distort images sent to Earth.

From Earth, computer processors at the Space Telescope Science Institute in Baltimore made adjustments to the orbiting telescope. They managed to remove some of Hubble's blurriness, but the images were still hazy. While Hubble circled Earth, engineers looked for ways to further correct the problem and redeem themselves.

One of the engineers pondering the situation was James A. Crocker. One night in 1993, while in Munich, Germany, Crocker stepped into a hotel room shower. "The shower head was on a bar," he explained later. "It ran up and down, and folded up." In the shower's simple construction, Crocker saw a way of saving Hubble.

Back home in Baltimore, Crocker raided his son's set of Ramagons, a toy construction set. Using plastic foam and pieces of Ramagon, he made a model of an oddly shaped device – a Swiss Army knife–like contraption with folding arms and twelve corrective mirrors. Crocker figured a full-scale version of the model just might correct the giant telescope's fuzziness if could be installed in the orbiting Hubble's belly.

Crocker pitched the idea to NASA. The device offered a glimmer of hope. Built to Hubble dimensions and called **COSTAR** (Corrective Optics Space Telescope Axial Replacement), the contraption weighed twice as much as a fridge.

In the laboratory, COSTAR worked like it should, automatically unfolding its octopus arms upon command. But would it work in space?

In December 1993, seven astronauts blasted into space aboard Space Shuttle Endeavor carrying COSTAR, an array of instruments, and two hundred custom-made tools. The six man, one woman crew had trained eleven months for the job. But even John Bahcall, an astrophysicist involved in the project, had doubts about the mission's success. "If they bring the repair mission off, it will be the equivalent of a modern-day miracle," he said. "I'll be there cheering and praying."

During the eleven-day flight, two teams of astronauts spacewalked around and through the 13 meter (43 ft.) long Hubble. Like delicate surgeons, they pulled out instruments to make room for COSTAR. Then they opened the phone-booth sized container that stored the device. The astronauts lifted it out, slid it into the empty slot, and screwed it into place on the telescope. To further correct the situation, they also installed a Wide Field and Planetary Camera.

The corrective surgery greatly improved Hubble's performance and reliability. From an idea spawned in a hotel room shower to fully installed device, COSTAR was the miracle that brought crisp views of distant galaxies back to Earth.

MORE AHA MOMENTS

Friedrich Kekulé (1865)
BENZENE'S MOLECULAR STRUCTURE

For years, Friedrich Kekulé, a German chemist, struggled to unlock the secrets of benzene, an organic compound with unique properties. In chemical reactions, other organic chemicals combined in predictable ways. But benzene was a renegade with odd and surprising properties. Its molecular structure was a mystery.

One evening, Kekulé fell asleep in front of a flickering fire. He began to dream. In his dream, atoms danced in mid-air. Some atoms linked up with others to form pairs. Some of the pairs joined other pairs. Chains of atoms joined other chains. The chains twisted and turned like snakes. Suddenly one of the snakes formed a circle, its head chasing its own tail. The head grabbed the tail, and the snake whirled around and around.

Kekulé awoke with a start, dazed by the still-fresh dream. He realized that the dancing snakes were the solution to the benzene problem. Rather than lining up in chains like other compounds, benzene's atoms formed a circle. That was the only way to explain benzene's peculiar and unpredictable ways.

Kekulé's vivid dream revolutionized chemistry, giving us new understandings of chemicals and the ways that they combine.

Richard & Betty James (1943)
SLINKY

Richard James, an American engineer, was assigned to correct an instrument problem on a ship when a heavy spring accidentally fell off a high shelf. Instead of crashing to the floor, the spring uncoiled and then gracefully flipped and slithered its way down. The accident inspired James and his wife, Betty, to make a toy out of a coiled-up spring. Called Slinky because of its unique flopping action, the toy has been a best seller since first hitting stores in November 1945.

George De Mestral (1948)
VELCRO

On a hike through the woods, annoying burrs stuck to Swiss engineer George de Mestral's clothing, forcing him to stop and pry them off. What made them so difficult to remove? Closer examination showed that the burrs had hook-like arms that locked into the open weave of his clothing. The discovery led de Mestral to invent a hook-and-loop fastener of his own. Today his invention – Velcro – can be found on everything from clothing and lunch bags to space suits and spacecraft.

Martin Cooper (1973)
CELL PHONE

At a time when telephones were stationary devices, Martin Cooper caught a glimpse of the future while watching a televised episode of Star Trek. In the show, Captain James T. Kirk used a Communicator, a fictional, handheld device the size of a deck of cards.

Inspired by the show, Cooper, head of Motorola's communications systems division, assembled a team to develop a product that worked the same way. Within 90 days, they had a prototype ready. It was the size of a brick and at 1.1 kilograms (2.5 lb.), just as heavy. To prove its mobile capabilities, Cooper invited journalists to meet him on Sixth Avenue in New York City. His first call was to Joel Engel, his chief competitor at AT&T, another communications company.

"Joel, this is Marty," he said brashly. "I'm calling you from a cell phone, a real handheld portable cell phone." Then he handed the phone to some of the reporters so they could make on-the-spot calls, too.

The stunt created a wave of publicity, but it took years of further development to miniaturize the device and set up an infrastructure of transmitting towers that would make the cell phone the communication wonder that it is today.

Hugh McCrory (1995)
HAIR PILLOWS

An otter swimming through an oil slick triggered Phil McCrory's brainstorm. McCrory, a hairdresser from Huntsville, Alabama, was watching a television program about oil pollution. One scene showed an otter swimming through the polluted area. McCrory was surprised to see how much oil collected on the otter's fur. Would human hair do the same thing?

As a hairdresser, McCrory had plenty of human hair on hand to test. His experiments showed that oil stuck to hair rather well.

In 1995, McCrory patented a pollution-fighting invention – "hair pillows". When tossed into an oil slick, the pillows absorb oil – basically, the oil sticks to the hair. When pillows are pulled out of the slick, they bring polluting oil along with them. Once the oil is squeezed out and collected, the pillows can be reused to clean up more of the mess.

MORE ACCIDENTAL DISCOVERIES

SUNSCREENS WERE INTRODUCED DURING WORLD WAR II TO PROTECT AMERICAN SOLDIERS WHO WERE STATIONED IN THE PACIFIC FROM SEVERE EXPOSURE TO THE SUN, BUT IT TOOK YEARS BEFORE DOCTORS REALIZED SUNSCREENS COULD ALSO BE USED BY SUNBATHERS TO PREVENT SUNBURN.

CHAPTER 2

FORTUNATE FUMBLES

Smashed bottles, jarring jolts of electricity, spilled chemicals, machines running amuck! Can such accidents ever lead to great breakthroughs?

Consider the case of the factory worker at the Proctor & Gamble Company in Cincinnati, Ohio. In 1878, he left for his lunch break in such a hurry that he forgot to turn off his soap-making machine, leaving it to churn for a much longer period that it should. Instead of reporting his error, the man packaged the batch of soap and sent the bars to customers thinking that no one would be the wiser. He was wrong. The soap had an unusual quality. Rather than sinking to the bottom of the tub, air bubbles trapped in the bar of soap caused it to float.

Customers loved the new soap since it meant no more fishing around in a sink or tub for a bar that had sunk to the bottom. When orders for the new product flooded the company, Proctor and Gamble discovered the worker's error and started mass producing Ivory Soap, "the soap that floats".

In this chapter, unplanned fumbles like this offer opportunities for breakthroughs that otherwise might be missed.

Pieter Van Muschenbroeck – 1746

A JAR OF ELECTRICTY

Have you ever walked across a carpeted floor and then touched a doorknob or a friend? The small shock you might receive is the discharge of static electricity. It is created by friction. When you walk across a carpet, your body becomes electrically charged. When you touch a conductor of electricity – a piece of metal for example, or another person – the charge can transfer to the object.

To produce static electricity hundreds of years ago, it was popular to use hand-cranked friction machines known as electrostatic generators. The problem with electrostatic generators was that they couldn't store electricity for use later. Once discharged by a touch, the charge was gone. To produce another charge, the machine had to be cranked again.

In 1746, a professor at the University of Leyden in Holland tried to make a device to hold electricity. Professor Pieter van Muschenbroeck and two assistants thought that they would be able to capture electrical charges if they surrounded an electrified object with a non-conductor such as glass.

To try it out, they hooked up an electrostatic generator to a brass chain that they dangled inside a glass jar. When they cranked the generator, electricity flowed down the chain to the jar. But when they touched a conductor to the jar, nothing happened. The electricity, it seemed, had disappeared.

Disappointed, the professor tried another approach. This time he filled the jar with water. Once more, he hooked up the chain to the machine and cranked the handle. Nothing. Again it seemed they had failed.

Discouraged, they began to dismantle the machine. An assistant held the jar with one hand while the water inside was still connected to the generator. With his other hand, he grabbed the wire to remove it from the machine. A jolt of electricity surged through his body, paralyzing his arms and legs.

The jar of water had indeed stored electricity, but it took an accident to prove the point. By touching the wire and the jar at the same time, the man acted as a shortcut for the electrical charge. After a few hours, he recovered, but the lesson lived on.

The apparatus became known as a Leyden jar. With it, strong currents could be stored for a long time, enabling scientists to discover new properties and benefits of electricity.

DID YOU KNOW?

MODERN-DAY ELECTROSTATIC GENERATORS CAN COLLECT HUGE CHARGES. THE VAN DE GRAFF GENERATOR, A DEVICE THAT USES A CONVEYOR BELT TO CARRY AN ELECTRICAL CHARGE TO A HOLLOW BALL, CAN DELIVER A WHOPPING 5 MILLION VOLTS OF ELECTRICITY.

Charles Goodyear – 1839
CHARRED RUBBER

Charles Goodyear's fascination with rubber started quite by accident. One day in 1834, Goodyear, an unemployed salesman, noticed a rubber life preserver on display in the window of a New York shop. He examined it carefully and found a defect in the product. He raced home, designed a new valve for the preserver to remedy the problem, and a few days later returned to the shop.

Goodyear was sure the manufacturer would buy his improved model. He was wrong. The problem with the life preserver, he was told, was not the valve. It was the rubber. Despite all its wonderful elastic properties, rubber had drawbacks, too. In cold weather it became so brittle it shattered into pieces; in hot weather it became sticky and foul-smelling. Find a way around those problems, the manufacturer told Goodyear, and his company would pay a fortune for the secret.

From then on, Charles Goodyear was hooked on rubber. With the little money he possessed, Goodyear bought big chunks of the stuff. He shredded it into tiny bits, loaded the pieces into pans, added chemicals, then stirred and heated the mixtures over his kitchen stove. He nailed samples of rubber all around his house to see what effect temperatures had on them.

For five years, Goodyear devoted himself to his experiments. He grew tired, pale, and sickly. Gradually his house emptied as, one at a time, he sold his household goods to buy food and clothing for his family and, of course, more rubber for his experiments.

Despite all the failures, Goodyear never gave up hope. Then unexpectedly in 1839, a small accident changed his life. He was experimenting with a mixture of rubber, sulfur, and white lead. As he stirred the batch, a bit of it splashed onto the hot stove. Instead of melting as Goodyear expected, it sizzled and charred around the edges. Curious about this strange reaction, Goodyear dropped another glob onto the stove. This time he noticed a thin rim of rubber between the charred edge and the rest of the material. The rim was flexible and moldable like rubber, but it didn't become brittle when cold or sticky when warm. Goodyear named his discovery "vulcanized rubber" after Vulcan, the Roman god of fire.

Although Goodyear had stumbled upon the right combination of chemicals to make vulcanized rubber, a question remained. How much heat was necessary? For the next five years, Goodyear experimented constantly in the family kitchen, trying to find the answer. His health worsened and his family became poorer as he struggled to finance his work.

Perhaps the ridicule Goodyear endured was worse than poor health and poverty. No one took him seriously. After all, he had failed repeatedly before. Now people laughed when he claimed that heat – the thing that made rubber sticky – was also necessary to cure its stickiness.

Finally, after ten years of struggle and misfortune, Goodyear discovered a winning formula. Pressurized steam applied for four to six hours at temperatures around 130°C (270°F) stabilized the mixture.

In 1844, Goodyear received a patent for his process, but even then he did not gain the wealth or respect he deserved. Other people claimed that they had invented vulcanized rubber first.

Goodyear continued to experiment on his own, developing many new uses for rubber. He sold his secrets to manufacturers. While they reaped huge profits, Charles Goodyear remained poor most of his life. When he died in 1860, he was $200,000 in debt.

Horace Wells – 1844

JUST ONE WHIFF

The hall in Hartford, Connecticut, was packed to overflowing. In the audience sat two friends – Samuel Cooley and Horace Wells, a dentist. Both young men were in for a surprise. Cooley didn't suspect that he was about to be the main source of entertainment. Wells didn't realize that he was about to change medical history.

When the speaker called for volunteers to participate in an experiment, Cooley strutted to the front of the audience. Would he sniff a little gas from a container to demonstrate its effects to the others? Not one to back down from a challenge, Cooley agreed.

The gas was nitrous oxide. Nitrous oxide had recently been discovered, and its unusual effects were a source of fun at gatherings like this. A good sniff of the gas usually turned unsuspecting subjects into giggling, laughing fools, much to the amusement of those in the audience. The effect led many to call nitrous oxide by another name – laughing gas.

Cooley inhaled deeply and broke into hysterical laughter. Then, as sometimes happened with nitrous oxide, Cooley's mood changed. He became violent. He scuffled with others and tried to pick fights with them. He tripped, fell heavily, and struggled to get up again. Momentarily stunned, he wandered back to his seat beside Wells.

Other volunteers were called forward and the demonstration continued. Someone glanced back at Cooley and noticed a pool of blood under his seat. The fall had gashed Cooley's leg and he was bleeding profusely. When informed of the injury, Cooley was surprised. He'd felt no pain at all.

Being a dentist, Wells realized the importance of this event. If nitrous oxide dulled a person's senses, perhaps it would kill pain during surgery. Wells wasted no time trying to prove his theory. A decaying molar caused him pain, so he gathered a few witnesses and asked a colleague to remove it. Before the dentist went to work, Wells inhaled nitrous oxide and quickly lost consciousness. While he was out, his friend extracted the tooth. Just as he predicted, Wells felt no pain at all.

Encouraged by his own experience, Wells arranged a bigger demonstration, this time in front of doctors and dentists at the Massachusetts General Hospital in Boston. He gave a willing patient a dose of nitrous oxide, then began removing one of the patient's teeth. To Wells' surprise, the patient screamed and howled in pain. In his nervousness, Wells had started the procedure before the gas had taken hold. The spectators booed and hissed, forcing Wells to leave in disgrace, his reputation ruined.

Wells soon gave up his dental practice, but his demonstrations aroused the interest of others who continued to experiment with nitrous oxide and other painkillers. Today, thanks to Wells and these medical pioneers going to the dentist is no longer the painful experience it once was.

Alfred Nobel – 1875

SPILLED LIQUID

Alfred Nobel had a personal interest in explosives. In 1864, his younger brother, Emil, had been killed in a tragic accident with nitroglycerine at the family explosives factory in Sweden.

In those days, nitroglycerine was widely used for blasting rocks in mines and quarries. But this highly unstable liquid often exploded unexpectedly – all it took was a slight jiggle of its container.

Impacted by his brother's death, Alfred Nobel looked for safer ways to use nitroglycerine. He found that if nitroglycerine was mixed with a porous white powder called *kieselguhr*, it could be rolled into sticks that could be carried safely. He called his new explosive dynamite.

With Nobel's discovery, a powerful explosive force was locked in a convenient, safe form that could be used in all areas of construction, from building roads to blasting tunnels. Driven by his success, Alfred Nobel opened up dynamite factories across Europe and became a wealthy man.

Always on the lookout for ways to improve the product, Nobel continued his research. One day in 1875, while experimenting with nitroglycerine,

Nobel accidentally cut his finger. Quickly he reached for a bottle of collodion and dabbed some on his finger. Collodion – a thick, sticky liquid – was often used on cuts because it dried quickly to form an elastic "skin" that sealed the wound.

As Nobel continued to work with nitroglycerine, a little of it dropped on the collodion. The collodion changed appearance. A gummy new substance formed.

Acting on a hunch, Nobel tried other experiments with collodion. He found that by heating mixtures of nitroglycerine with finely divided collodion, a transparent, jelly-like explosive product formed. It was even more powerful than dynamite.

Alfred Nobel called his discovery blasting gelatin. Because of his lucky accident, it is often said that "blasting gelatin was born on a man's finger and not in a test tube."

DID YOU KNOW?

BESIDES DYNAMITE AND BLASTING GELATIN, ALFRED NOBEL HELD 354 OTHER PATENTS FOR DISCOVERIES AND INVENTIONS. IN 1895, NOBEL WROTE HIS LAST WILL, LEAVING MUCH OF HIS VAST WEALTH TO THE ESTABLISHMENT OF THE NOBEL PRIZE. HE DECREED THAT HIS FORTUNE SHOULD BE DIVIDED INTO FIVE PARTS AND USED FOR ANNUAL PRIZES IN PHYSIOLOGY OR MEDICINE, PHYSICS, CHEMISTRY, LITERATURE, AND PEACE "TO THOSE WHO, DURING THE PRECEDING YEAR, SHALL HAVE CONFERRED THE GREATEST BENEFIT ON MANKIND."

Edouard Benedictus – 1903
UNBREAKABLE GLASS

As French chemist Edouard Benedictus climbed a ladder to retrieve chemicals on a high shelf, his hand slipped, knocking a glass flask to the floor. The flask ricocheted off the hard surface, but instead of shattering into shards, it cracked and kept its original shape.

Fascinated by the odd behavior, Benedictus examined the flask. A thin film coated the inside. It seemed to be holding the broken pieces together.

With a little investigating, Benedictus identified the filmy substance. The flask once contained a solution of collodion. Over time, the sticky liquid had evaporated, leaving a thin residue behind.

Benedictus labeled the flask, returned it to the shelf, and continued with his original research. Weeks later, he read a newspaper story about a young girl who had been badly cut by flying glass in an automobile accident. He remembered the non-shattering flask. Could the filmy coating prevent accidents like this?

Benedictus began a new series of experiments. He spread collodion between sheets of glass and used a press to bond the sheets together. Tests showed that the glass resisted shattering. Even when it splintered upon impact, the pieces stayed together.

Eventually, Edouard Benedictus developed safety glass, a type of glass that stays together even when smashed with a hammer. Today we use safety glass in many places – car windshields, windows and doors of public buildings, even in goggles for machinists.

MORE ACCIDENTAL DISCOVERIES

IN 1863, PRINTER JOHN WESLEY HYATT DISCOVERED THAT A SPILL OF COLLODION IN A MEDICINE CABINET HAD HARDENED INTO A THIN, CLEAR SHEET. INTRIGUED, HE MIXED COLLODION WITH SAWDUST AND PAPER, HOPING TO PRODUCE A HARD, DURABLE SUBSTITUTE FOR IVORY. HE ENDED UP WITH A NEW PRODUCT, CELLULOID, THE FIRST OF MANY PLASTICS.

James W. Christy – 1978

AN UNEXPLAINED BUMP

In June of 1978, astronomer James Christy followed a familiar routine. He took a recently snapped photographic plate of the night sky, placed it into a machine called a Star Scan, and turned it on. Christy had scanned dozens of other photographs in the same way. At first glance, this one looked no different from the others. The stars were like pinpoints of light. To no surprise, the dwarf planet Pluto looked like a hazy ball.

But something caught Christy's attention. There was a bulge on Pluto. A blurry bump. Pluto looked stretched, elongated. Was there a smudge on the photographic plate? Or had there been some unexpected movement at the moment the picture was taken? Whatever the reason, the photograph appeared defective. Christy decided to scrap it.

Just then the Star Scan machine flickered and died. Christy called for help. While a technician worked on the machine, Christy stayed in the room, ready to give assistance if necessary. In the hour it took to complete the repair, Christy idly studied the ruined photograph. The hazy bump bothered him. Could there be more to it than he had first thought?

Christy went to the archives, the room where earlier photographs were stored. He found one marked "Pluto image. Elongated. Plate no good. Reject." In this photograph, Pluto had the same bumpy look. Digging further, Christy found six other rejected photographs taken between 1965 and 1970.

Each one had been discarded for the same reason. Each one showed the same bulge.

The rejected photographs, it turned out, were not defective at all. The curious breakdown of the Star Scan machine gave Christy a chance to take a second look, wonder, ask questions, and make an important discovery. The blurry bulge turned out to be a moon, the first ever discovered for Pluto. Christy named it Charon, after his wife, Charlene.

DID YOU KNOW?

THE COMBINATION OF LUCK AND CLOSE OBSERVATION PLAYED A ROLE IN OTHER DISCOVERIES IN ASTRONOMY.

IN 1932, KARL JANSKY, AN ENGINEER FOR BELL LABORATORIES, WAS TRYING TO FIND SOURCES OF NOISES THAT INTERFERED WITH LONG-DISTANCE RADIO SIGNALS. HE RIGGED UP A WEIRD LOOKING ANTENNA SYSTEM, MOUNTED IT ON A JUNKED MODEL T FORD, AND DROVE AROUND LOOKING FOR RADIO INTERFERENCE. TO HIS SURPRISE, HE DETECTED A PERSISTENT HISS COMING FROM THE MILKY WAY GALAXY. JANSKY WAS THE FIRST PERSON TO DISCOVER THAT MANY BODIES IN THE UNIVERSE EMIT RADIO SIGNALS. JANSKY ALSO CLAIMED ANOTHER FIRST: HIS RICKETY ANTENNA-ON-WHEELS WAS THE WORLD'S FIRST EVER RADIO TELESCOPE.

Frank Etscorn – 1986

LIQUID NAUSEA

At some point on an otherwise ordinary day, behavioral psychologist Frank Etscorn stumbled and tripped. He had been walking across his laboratory in the basement of the New Mexico Institute of Mining and Technology, carrying an open vial containing a brown liquid.

Etscorn was studying sugar dependency in rats. The liquid was a nicotine extract, a nausea-inducing substance found in tobacco that Etscorn planned to use on rats to see if it reduced their craving for sweets. But that day, Etscom tripped. The liquid sloshed on to his arm, giving him – not the rats – a highly concentrated dose of nicotine.

"I wiped it off and didn't pay attention," he told a reporter for People Magazine later. "But after about 15 minutes I felt nauseated."

The experience sidetracked Etscorn, steering him into a new area of research. "I had a great idea," he said. "This would be a great way to get nicotine into the skin. Almost immediately, I also realized this could be a way for people to stop smoking."

What Etscorn envisioned was a slap-on patch similar to the ones already being used to control motion sickness. By giving the wearer of the patch steadily reduced doses of nicotine over a long period, Etscorn figured smokers could be weaned off their addictive habit.

Having the idea was one thing. Developing an effective, reliable, cost-effective patch proved more complicated. For months, Etscorn, a non-smoker, experimented on himself. He swabbed different concentrations of nicotine on to his body and charted their nausea-producing effects. Then he convinced his brother, John, a long time smoker, to be a subject, too.

When John visited his brother in 1981, Frank tested liquid nicotine on him. "I flopped him on our kitchen table and smeared some of it on the hollow of his neck." In minutes, nicotine seeped into John's body and he felt the results.

By 1986, Frank Etscorn had a workable product and a patent to protect it. Called *Harbitol*, it was the first nicotine patch of its kind.

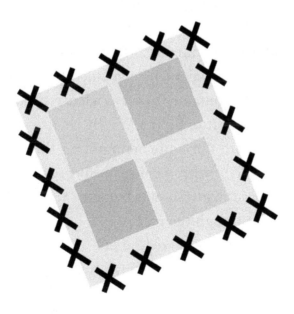

MORE FORTUNATE FUMBLES

Berthold Schwartz (about 1350)
GUNPOWDER

Some say the Chinese should get credit for coming up with the idea almost two thousand years ago. Other accounts mention a monk named Berthold Schwartz who lived in Freiburg, Germany, in the fourteenth century.

One day, as the story goes, Berthold was making a medicinal mix from three readily available ingredients – sulfur, charcoal, and sodium nitrite. After grinding each of the chemicals into fine powders, he combined them and placed a stone on top of the container to act as a loose-fitting lid. Later Berthold struck a flint to light a fire. By chance a few sparks flew into the container, igniting the mixture. It exploded with a bang, hurling the stone upward with so much force that it smashed through the roof.

Fascinated, the monk mixed other batches of the explosive combination. He even stuffed some into a hollow tube that was closed at one end. After putting a stone ball on top of the open end, Berthold touched a flame to a tiny hole in the side of the sealed tube. The mixture exploded, propelling the stone ball with great force. Eureka! The first gun. Or so the story goes.

Benjamin Franklin (1750)
LIGHTNING ROD

Benjamin Franklin – American politician, inventor, and a man with an odd sense of humor – set out to dazzle his friends. He invited them to a Christmas party in his home. Once everyone had gathered, Franklin brought out a live turkey, the main course for dinner. Electricity was a fairly new source of energy at the time, and Franklin hoped to impress his friends by killing the turkey with it.

But things went wrong. Franklin accidentally touched one of the connectors and was sent flying. He survived, but the loud bang and simultaneous flash of light produced by the discharge reminded him of lightning. The experience led Franklin to his most famous and dangerous investigation – launching a kite in a thunderstorm to test the properties of lightning. Eventually, Franklin invented the lightning rod, a safety device that diverts lightning to the ground and is still used in buildings today.

James Hargreaves (1764)
SPINNING JENNY

James Hargreaves, a weaver and carpenter, was watching someone use a spinning wheel when suddenly the spinning wheel fell over. As it lay on its side, the wheel continued to turn and so did the spindle or shaft on which the thread was wound. But instead of being in a horizontal position parallel to the ground, the spindle was now in a vertical position and sticking up in the air. The accident gave Hargreaves an idea. If the spindle was set in an upright position, more than one could be used at a time. Several threads could be spun at once.

By 1764, Hargreaves had developed a machine with upright spindles that could spin five times faster than by hand. He called it the "spinning jenny" after his daughter Jenny.

Thomas Edison (1877)
PHONOGRAPH

Thomas Edison, the famous American inventor, was working on a telephone, hoping to find a way to improve it. When he spoke into the receiver, a needle fastened to the diaphragm vibrated, accidentally pricking Edison's finger. That incident made Edison stop and wonder. If sound vibrations were passed along a needle, could they etch lines into a soft material? Would a needle passing over those same lines repeat the original sound? Inspired by the accident, Edison went on to invent the world's first "talking machine", a device which duplicated sound by passing a vibrating needle over grooves that had been cut into tin foil. Edison's invention was the forerunner of the sophisticated sound-recording devices used today.

Wilson Greatbatch (1956)
IMPLANTABLE PACEMAKER

While building an oscillator to record the sound of a heartbeat, American inventor Wilson Greatbatch mistakenly used the wrong type of transistor. Instead of recording sounds, the device produced electrical impulses. The impulses mimicked the rhythm of a human heartbeat, and Greatbatch immediately realized the value of his error. "I stared at that the thing in disbelief, thinking this was exactly the properties of a pacemaker."

Pacemakers of the time were bulky, external devices that delivered electrical shocks to stimulate and regulate a patient's heartbeat. With his transistor-operated instrument, Greatbatch saw a way to miniaturize the pacemaker and make it one that could be surgically placed inside the body. In May 1958, he demonstrated the product's usefulness by implanting it in a dog. By 1960, Greatbatch's invention was being used in human subjects, giving added life and mobility to people with heart trouble.

Ichiro Endo (1977)
INKJET PRINTER

While looking for alternative methods of printing for the Canon Company in Japan, engineer Ichiro Endo accidentally dropped a hot soldering iron on to a syringe filled with ink. Moments later, Endo noticed ink seeping from the syringe nozzle. He realized that heat had boiled the ink and forced bubbles to shoot out the end. Curious about the phenomenon, Endo recreated the incident and filmed it with a high-speed camera. Within 3 days of the accident, Endo and his team of engineers had built a working prototype of a new device – the ink-jet printer still used today.

CHAPTER 3

OPPORTUNITY KNOCKS

Coincidences and chance events happen to us all the time. Usually we don't notice. They don't alter our lives. But to an observant person, to one on the brink of change or facing a problem, chance occurrences can be a source of inspiration. Consider the case of Samuel Morse, an American portrait painter.

When Morse boarded a ship in France on October 1, 1832, he was a man at the peak of his artistic career. On the second evening of the voyage across the Atlantic, Morse and a few other passengers gathered in the dining room. A discussion began. The topic: electricity.

One of the passengers described how to make an electromagnet by wrapping coils of insulated wire around a metal rod and then connecting the wires to a battery. The greater the number of coils, he told the others, the greater the electromagnet's power.

One of the passengers asked a question. "If you use more wire, won't you slow the electricity? Won't it take longer for the electricity to travel?"

"No," the knowledgeable man explained. "Electricity passes instantly over any length of wire, even if it is a mile long."

Hours later, while most passengers slept, Samuel Morse lay awake in bed, still thinking about the man's words. *Electricity passes instantly over any length of wire.*

Suppose . . . Morse wondered . . . suppose a message could be sent along with the electrical current. Would the message be carried instantly, too?

By morning, Morse had reached a life-changing decision. He abandoned his artistic career and devoted his time and fortune to finding a way of sending messages at the speed of electricity.

The simple conversation aboard the ship spawned a brainwave. After years of trial and error, frustration and failure, Morse succeeded in launching a revolutionary invention that used electromagnets to send rapid-fire messages. Morse's invention – the telegraph – and the code of dots and dashes that Morse also invented, changed the world of communication, bringing news to the masses almost as soon as it happened.

In this chapter, unexpected twists of time and place – those coincidences, chance encounters, and casual observations that many of us ignore – become ripe opportunities for discovery and invention instead.

René Theophile Laennec – 1816

CHILDREN'S GAME, DOCTOR'S TOOL

If you put your ear against someone's chest, you might hear the person's heart beating. Until 1816, that's how doctors examined their patients. Then René Laennec, a Paris doctor, stumbled upon something better.

On his way to visit a patient one day, Dr. Laennec walked along the streets deep in thought. His patient, a young woman with heart disease, was extremely overweight. Laennec worried that he might not be able to hear her heart beating.

The sound of laughter interrupted his thoughts. Several children were playing a game on a pile of old lumber. While one child pressed his ear against one end of a long wooden beam, another child tapped the other end. The children squealed with delight when they heard the sound travel through the length of board.

Later, as he was about to examine his patient at her home, Laennec recalled the children's game. Why, of course! He took a sheet of paper and rolled it into a tube. When he pressed one end of the tube against his patient's chest and listened at the other end, he heard the movements of her heart clearly.

Laennec experimented with different materials for his listening device. Being an expert wood turner, he produced a cylinder of wood about 30 centimeters (12 inches) long. It was hollow in the center and had adjustable cups at each end. When asked to give his invention a name, René Laennec combined two Greek words – *stethos*, meaning 'chest', and *skopos*, meaning 'observer', and created a new word – *stethoscope*.

Today doctors use a variation of this instrument that was inspired by a children's game. It may look different than Dr Laennec's original invention, but the stethoscope remains the simplest way for a doctor to listen to the beating heart.

MORE ACCIDENTAL DISCOVERIES

CHILDREN AT PLAY HAVE BEEN INVOLVED IN OTHER IMPORTANT DISCOVERIES. IN 1812, WHILE PLAYING ALONG A ROCKY BEACH IN ENGLAND, TWELVE YEAR-OLD MARY ANNING SPOTTED AN UNUSUAL FEATURE IN THE LIMESTONE CLIFFS – A NARROW SKULL WITH ROWS OF POINTED TEETH. THE SKULL TURNED OUT TO BELONG TO A LONG-EXTINCT SEA REPTILE WHICH SCIENTISTS DUBBED ICHTHYOSAUR – "FISH-LIZARD".

William Beaumont – 1822

HOLE IN THE STOMACH

William Beaumont, an army surgeon, was part of a peacekeeping force the United States Army stationed on Mackinac Island in Lake Huron, near the shores of present-day Michigan. The ambitious Dr. Beaumont felt that his talents were wasted on the calm little island. Even at the peak of the fur trading season, few incidents called for a surgeon's skills. But then along came Alexis St. Martin and the doctor's peaceful life changed forever.

Alexis St. Martin was a carefree, adventure-loving, French Canadian youth. He had been hired as a voyageur by the American Fur Trading Company to transport furs by canoe through the wilderness of North America. With the fur trading season drawing to a close, St. Martin, like other voyageurs returned to the small village on Mackinac Island that served as the company's headquarters.

On June 6, 1822, St. Martin was relaxing with other men in the company store. Nearby a drunken voyageur toyed with his newly purchased gun. Suddenly the gun went off. A full load of buckshot and powder ripped into Alexis St. Martin's body just below his chest.

Dr. Beaumont raced to the youth's side, amazed that the voyageur was still alive. A huge hole as large as a human hand penetrated St. Martin's abdomen. Part of his stomach and part of one lung hung out of the cavity.

Beaumont cleaned the wound and applied a dressing. He fully expected the voyageur to die. But Alexis St. Martin did not. His wound healed in a peculiar way. Rather than settling back into the abdomen, his stomach attached itself to the chest wall. Scar tissue formed around his wound, but the hole remained open. A loose flap of stomach lining hung over it like a shade pulled over a window. By pushing aside the flap, Dr. Beaumont could see inside St Martin's stomach.

Beaumont recognized a rare opportunity that he couldn't pass. Curious about digestion and how the body processed food, he proposed a series of painless experiments to Alexis St. Martin. No longer having the strength or endurance to work as a voyageur, and dependent on the doctor for food and shelter, Alexis St. Martin agreed.

In one experiment, Beaumont tied tiny bits of food to silk threads and lowered them through the hole into St. Martin's stomach. Now and then he lifted them out, observed the state of digestion, and then returned them to the stomach. In other tests, Beaumont extracted, analyzed, and experimented with stomach juices. Once he even poked a thermometer through the hole to check the stomach's temperature.

For twelve years, Dr. Beaumont conducted experiment after experiment. He became a respected and admired authority across North America and Europe. His findings startled doctors and scientists, and led to the development of a new branch of science – nutrition, the study of food and how the body uses it.

But what of Alexis St. Martin? For fifty-eight years after the accident, St. Martin continued to be a medical wonder. He received dozens of invitations to appear before interested groups and show his now famous hole. The fees that he collected helped support his growing family.

DID YOU KNOW?

ALTHOUGH HIS METHODS WERE UNUSUAL, DR. BEAUMONT'S EXPERIMENTS MUST HAVE BEEN HARMLESS. ALEXIS ST. MARTIN LIVED TO THE RIPE OLD AGE OF EIGHTY-THREE, OUTLIVING THE DOCTOR BY TWENTY-SEVEN YEARS.

John and Allan McIntosh – 1835
FRUIT FROM A SINGLE TREE

Many varieties of apples are grown worldwide, but the undisputed favorite of many apple-eaters is the McIntosh Red. Millions of this crisp, juicy fruit are harvested each year. But the McIntosh Red had humble beginnings, and if it wasn't for a bit of luck this popular fruit might not be in kitchens today.

In 1811, a Scottish settler named John McIntosh moved to a homestead in Dundas County, Ontario, Canada. To prepare the land for farming, he cleared trees off his property. Hidden in the dense bush, John found a cluster of twenty young apple trees.

To the young farmer, finding the trees was as good as finding gold. Apples were a valuable commodity to pioneer settlers. The fruit added variety to an otherwise bland diet. Apples were versatile, too. They could be eaten straight off the tree in summer, stored in a cool place for the winter, cooked into cakes, pies, and other delicacies or even squeezed into refreshing juice or cider.

John uprooted the young trees and transplanted them closer to his home. The following season, he had an ample supply of tasty apples. However, one of the trees produced an especially sweet and delicious fruit. It quickly became a favorite with the whole family and with neighbors near and far.

Unfortunately, one tree could hardly produce enough fruit to satisfy a single family, let alone the entire neighborhood. Although John tried planting seeds taken from the fruit, the new trees did not bear the same apples. As John soon discovered, apple trees do not "breed true". That is, seeds from the fruit do not produce apples exactly like the parent tree.

For over twenty years, the tree continued to bear fruit, but never enough to satisfy everyone. Then one day in 1835, a wandering farm hand happened to hear the story of the single tree with the delicious fruit. He offered to help.

Taking a sharp knife, he cut a short twig from the apple tree, walked over to a young seedling tree, carved a small slit into its bark near the top, and inserted the twig. Using some twine, he wrapped it around the joint to keep it in place.

This procedure is called grafting. As the seedling grows, the twig grows as well. It becomes the top of the new tree and bears the same fruit as the original.

The method worked. Allan – John's oldest son – began traveling around the area selling branches of the tree to other farmers. With each sale, he showed them how to graft the branches to their own apple seedlings.

The original tree on John McIntosh's farm died in 1910. By then the McIntosh Red could be found all over Canada and the United States. Today offspring of the original tree grow in far off places all around the world.

David Fife – 1843

SEEDS IN THE MAIL

David Fife was an unusual man. Instead of planting a single type of grain like his neighbors in Peterborough, Ontario, Canada, Fife planted many different varieties each year. By dividing his farm into small experimental plots, each one growing a different type of wheat, he hoped to find the hardiest, healthiest, and most productive strain.

In 1843, Fife received some grain seeds in the mail from a friend in Scotland. Fife's wife, Jane, was ill at the time so he could not plant the seeds right away. When he finally got around to sowing them, it was late in the season. Most of the other strains were growing by this time.

At first, the new wheat seemed doomed to failure. Out of all the seeds, only one sprouted. Fife was tempted to plow the single plant under and start over with another strain of wheat. But then he noticed something unusual about the plant. It had three stalks instead of just one, and it seemed to grow quickly. By mid-summer, it had caught up to the others. When most of the other strains weakened because of disease, this one remained healthy and strong.

The new plant ripened earlier than the others. Then, just as it was ready to harvest, one of the cows broke through the garden gate, trampling and eating every plant in sight. Jane spotted the cow from the kitchen window just as its tongue was about to wrap around the tender stalk of the new wheat. She ran into the yard, waving her apron high in the air, yelling at the top of her lungs. The cow wandered away, and the wheat was saved.

From this single plant, Fife obtained more seeds. The new wheat was especially hardy, enabling it to survive diseases and cold temperatures that killed other strains. It also required a shorter growing season which meant that it could be planted later in the spring and harvested earlier in the fall. As well as producing a high yield of grain, flour from this wheat made delicious breads and pastries.

This new variety of wheat was called Red Fife. It proved to be ideal for the rugged prairies of Canada, the northern United States, and other areas around the world.

1848

PERSONALITY SWITCH

One day Phineas Gage was a cheerful, ambitious man. The next he was snarly, lazy, obnoxious, and the world of medicine was never quite the same.

Twenty-five year-old Phineas Gage was a track layer for the Vermont Railway Company. Part of his job was to blast away rocks to prepare the rail bed for new track. One day Gage poured gunpowder into holes that had been drilled into the rock. To pack the gunpowder, he used a long, pointed iron rod called a tamping iron.

Usually, this was a fairly safe activity. That day, it wasn't. The tamping iron struck a nearby rock, creating a spark. The gunpowder ignited and blew the rod right at Gage's head. The pointed end hit just below his left eye, ripped through his brain, and punched a hole in the top of his skull.

The tamping rod landed 45 meters (148 ft.) away and Phineas Gage was hurled to the ground. Blood poured from the wound. His hands and legs twitched. His co-workers figured he was dead, but in a few minutes Gage sat up, dazed and bloody, yet somehow still alive and able to speak. He was carried in a sitting position to a nearby town where local doctors treated him.

Gage's chances for survival were slim. He had lost a lot of blood, and in a matter of days, the wound became infected. A local cabinet maker was hired to build a coffin. To everyone's surprise, however, Gage made an amazing recovery. In just three weeks he was up and about, anxious to return to the life he once had.

But Gage was not entirely well. From the moment of the accident, he was a totally different person. His character changed. Instead of being the happy, responsible man he once was, Gage became abusive, argumentative, and unreliable. One person described him as a man with "the strength of an ox and an evil temper to match." He lost his job with the railway company, and eventually joined a traveling circus as a sideshow curiosity.

The incident attracted the attention of doctors around the country. Phineas Gage's accident proved that the entire brain was not necessary for life, but the sudden change in his personality raised many questions. Was there an area of the brain that was solely responsible for character? Were different regions of the brain responsible for other functions such as language, speech or sense of smell?

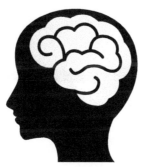

To find the answers, doctors examined and questioned Gage. They conducted tests and found examples of other people who had suffered head injuries. Slowly, over many years, pieces of the brain puzzle began to fall into place.

Doctors now know that the frontal cortex, the area of Gage's brain that was damaged, controls personality. They know, too, that other regions govern other functions. Today surgeons can even pinpoint these sites and do delicate operations to correct damaged areas, but in Phineas Gage's time little was known about the brain. It took a freak accident along a deserted stretch of railway to lead us down the path to discovery.

> **DID YOU KNOW?**
>
> AFTER HIS DEATH, THE TAMPING ROD AND THE SKULL OF PHINEAS GAGE WERE SENT TO THE MUSEUM AT HARVARD MEDICAL SCHOOL WHERE THEY ARE STILL ON DISPLAY. THE SKULL SHOWS THE HOLE CAUSED BY THE 6 KILOGRAM (13 LB.) TAMPING ROD WHILE A CAST OF HIS HEAD SHOWS THE SCAR THAT THE WOUND CAUSED.

Donald Johanson – 1974

A FLOOD OF BONES

As soon as Donald Johanson awoke on the morning of November 30, 1974, he sensed the day would be special. "I felt it was one of those days when you should press your luck," he said later, "one of those days when something terrific might happen."

Johanson was a paleoanthropologist, a scientist who searches for fossil evidence of the earliest humans. For weeks, he had been scouring a desert region in Ethiopia, Africa. He had found the occasional bone fragment, usually of some animal, but nothing substantial, nothing that gave him clues about human life long ago. He would search the area one last time, he told himself.

A heavy rain had recently swept through the region, causing a flash flood. Johanson walked along a gully created by the flood. He dodged overturned stones and picked his way along deep furrows formed by the rushing water. After walking for hours, he was ready to give up. He had found nothing unusual, no treasures of the past.

Suddenly, from the corner of his eye, he spotted a small object protruding from the eroded bank. It was a fossilized arm bone. Scattered nearby he found other bone fragments. Johanson quickly realized that the bones were humanlike and very old. He hurried back to the base camp to tell his colleagues.

Over the next three weeks, the team excavated the site and found dozens of bones. All of them belonged to a single individual: an adult female who had lived millions of years ago. Johanson called her Lucy after a popular Beatles song of the day titled *Lucy in the Sky with Diamonds*.

Nature had given Johanson a helping hand, but more than luck was involved in the discovery. Although the flash flood had churned the soil, bringing the long-hidden bones to the surface, it was Johanson's keen eye that recognized their true value. The discovery caused great excitement in scientific circles. Lucy proved to the oldest and most complete prehistoric human ancestor ever found until that time.

DID YOU KNOW?

IN 1851, QUARRYMEN DIGGING A CAVE IN THE NEANDER VALLEY REGION OF GERMANY FOUND A FOSSILIZED SKULL CAP, RIB BONES, AND OTHER SKELETAL PIECES. THE FOSSILS WERE OVER 300,000 OLD AND BELONGED TO AN EXTINCT SPECIES OF PRIMITIVE HUMANS WITH APE-LIKE FEATURES KNOWN AS NEANDERTHALS. THE ACCIDENTAL DISCOVERY OF THE NEANDERTHALS IS CREDITED WITH STARTING A NEW BRANCH OF SCIENCE – PALEOANTHROPOLOGY, THE STUDY OF THE ORIGINS OF THE HUMAN SPECIES.

MORE OPPORTUNITY KNOCKS

Elias Howe, Jr. (1839)
SEWING MACHINE

Elias Howe, Jr. worked as an assistant in a Boston machine shop. One day he overheard the owner and another worker argue. Elias tried to concentrate on his job, but as the voices grew louder, he couldn't help but listen to their conversation. "Why waste time on a knitting machine?" he heard the owner say. "Invent a sewing machine and you'll make a fortune."

To the penniless Elias – who earned only two or three dollars a week – these words seemed like magic. A fortune? Just for a machine that sewed?

These simple words fuelled Elias's imagination. For years he struggled, passing over one disappointment after another until at long last he achieved his goal and invented the first modern sewing machine. When he died in 1872, at the age of forty-eight, he had amassed an incredible fortune – over thirteen million dollars.

George Eastman (1874)
KODAK CAMERA

When young George Eastman set out on his vacation, he decided to take photographs of the trip. He gathered the cumbersome equipment used at the time: a bulky camera the size of a microwave oven, a heavy tripod, and assorted glass plates, trays, chemicals, and other supplies to develop the film. Picture-taking proved so difficult and expensive that Eastman canceled his trip and devoted all of his spare time to finding a way of making it simpler and more convenient. After four years of trial and error, he invented a small box camera that used lightweight film instead of heavy glass plates. His camera, the Kodak, revolutionized photography and made picture-taking available to everyone.

Christian K. Nelson (1920)
ICE CREAM BAR

Christian Nelson owned a candy and ice cream store in Onawa, Iowa. One day a young boy came into the store clutching a few coins and wanting a chocolate bar. After a few seconds, the boy changed his mind and ordered an ice cream sandwich. Before

Nelson could fill the order, the boy switched again. Make it a chocolate bar after all.

Why couldn't the boy have both tasty treats at the same time? Nelson thought. He experimented with chocolate and ice cream, hoping to find some way to make the chocolate stick to the surface of the ice cream. Eventually, he succeeded and created the first ever ice cream bar.

William F. Grimes (1954)
TEMPLE OF MITHRAS

After 57 days of intense bombing by Nazi Germany during World War II, large areas of the city of London, England, lay in ruins. Archeologist William F. Grimes, director of the London Museum, seized the moment to search through the rubble for the lost town of Londinium, a Roman settlement founded in 43 A.D. In 1954, along Walbrook Street, his team of archeologists discovered the ruins of the Temple of Mithras, built by the Romans between 240 and 250 A.D. Today marble sculptures and other treasures from the site are on display in the Museum of London.

Philippe Kahn (1997)
CELL PHONE CAMERA

During much of his wife's eighteen-hour labor, American entrepreneur Philippe Kahn sat at a nearby desk, cell phone, laptop, and digital camera at his side, ready to snap pictures of his first-born child. Internet technology was still in its early stages, and Kahn thought about the complicated steps ahead – downloading photos to his computer, posting them on a website, contacting relatives to tell them where to find the pictures . . .

With time on his hands, Kahn fiddled with the equipment, aiming to simplify the process. Kahn wrote a computer program, made a few trips to Radio Shack for supplies, and by the time his daughter Sophie made her appearance, he was armed and ready. His makeshift camera-phone-computer device streamlined the entire operation, allowing him to snap pictures and automatically post them on the Web.

Kahn's invention was the first of its kind and the forerunner to the much smaller, built-in cell phone camera carried by millions today.

CHAPTER 4

EXPERIMENTAL TWISTS

Think of the words "scientist" and "inventor". What comes to mind?

It's likely that you get an image of a person in a cluttered laboratory wearing a white lab coat and surrounded by bubbling mixtures. The person is tinkering with bottles, vials, and test tubes. As he or she pours one solution into another, the liquid froths and changes color. Clouds of choking smoke billow across the room. Clearly something has gone wrong.

This is the image often painted of scientists and inventors. It is a distorted image, but in one way at least, it is accurate.

Experiments do not always go as planned. Sometimes the results are unusual, unexpected, even disappointing.

These surprises aren't necessarily bad. Sometimes unforeseen kinks in an experiment force a scientist or inventor to pause, study the situation, and see unexpected possibilities. Sometimes a twist of fate during an experiment can be the key that unlocks the door to a totally new discovery.

Elihu Thomson – 1876

FLASH OF GREEN

No other science class quite matched Elihu Thomson's. While most teachers in the 1870s taught only by lecturing, Thomson's classes were filled with dynamic demonstrations. He challenged his students at Central High School in Philadelphia to think, question, and then experiment in the school laboratory, the only one of its kind in the United States.

As the young professor's popularity grew, he was frequently asked to be a speaker at public functions. In the fall of 1876, the Franklin Institute – a group of distinguished scientists – invited Elihu Thomson to give five winter lectures to its members.

Following his usual style, Thomson carefully prepared his material and included several demonstrations to highlight key points. The first four lectures went as planned. The fifth did not.

In the last lecture, Thomson wanted to show electricity in action. On a table in front of his audience, he placed a Leyden jar. Beside it, he had a hand-operated electrostatic generator and several copper wires.

First Thomson connected the generator to the Leyden jar. He cranked it vigorously, sending electrical charges to the jar where they were stored. In this way, he had a large source of electricity ready for use.

Thomson planned to cross two copper wires that led away from the Leyden jar. When the two wires made contact, a sudden spark would prove to the audience that electricity had passed from one wire to another.

The audience knew of Thomson's reputation for showmanship. No one wanted to miss the action. The hall was hushed, all eyes watching the equipment at the front. Thomson carefully brought the wires together. Instead of the spark that Thomson expected, a bright green flash sizzled across the points of contact. Startled but unfazed, Thomson picked up the two wires. They were solidly fused together.

Although Thomson knew something special had happened, he did not want to lose his audience. He dismissed the unusual incident with a casual remark and continued his lecture.

Later in his laboratory, Thomson toyed with the fused wires. How and why did this happen? Long years of experimentation followed. Eventually, Thomson devised a practical way of using electricity to melt metals and weld them together. Before long, dozens of industries from shipbuilding to toy manufacturing were using electric welding to join metal parts cleanly and simply.

Louis Pasteur – 1880

A WEAKENED STRAIN OF BACTERIA

Smallpox . . . scarlet fever . . . polio. Years ago these were common diseases. Today early vaccinations have made them almost non-existent. A bit of carelessness over a hundred years ago helped make them that way.

Louis Pasteur, a chemistry professor in Paris, believed that microscopic germs caused diseases. He thought that germs were outside the body, and diseases started when germs entered it. Not many people shared his belief, but Pasteur was determined to prove his ideas were right.

Around 1880, he began to study a contagious animal disease called chicken cholera. An associate sent him the head of a rooster that had died of the disease. Convinced that the rooster's blood contained disease-causing germs, Pasteur tried to isolate them.

First he prepared a bottle of broth from chicken gristle. He added a drop of the rooster's blood and placed the liquid in a warm place. After a few hours, Pasteur examined a drop of the mixture under his microscope. With food and warmth, the germs had multiplied. Hundreds swarmed in the culture.

Pasteur mixed a tiny drop of the culture with one chicken's food. Soon after eating the mixture, the chicken died. Pasteur tested the effects again and again. Each time a chicken died, adding proof to Pasteur's theory that disease started when germs entered a healthy body.

Fascinated by his experiments, Pasteur worked long and hard, convinced he was close to gathering the evidence he needed. His wife was concerned with his health, however. When she persuaded him to go on a well-deserved three-month vacation with his family, Pasteur left two assistants to look after the cholera cultures.

In Pasteur's absence, his overworked assistants figured they needed a holiday, too. For weeks, the cultures were left unattended. Many of the germs died and the strain weakened.

When he returned, Pasteur was angry but also curious. Would the weakened germs affect healthy chickens in the same way as before? Pasteur injected several chickens with the weakened cultures. Instead of dying, the chickens became only slightly ill and then recovered completely. This was new!

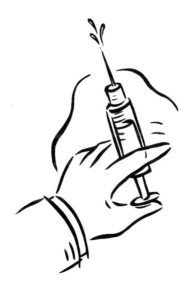

Pasteur experimented with two groups of chickens. One group received injections of the weakened cholera strain. The other group – the control group – didn't. Afterward, Pasteur injected both groups of chickens with a fresh batch of cholera. Then he waited.

Over the space of a few hours, the chickens in the control group died one by one while the chickens injected with the weakened strain remained healthy. Injecting animals with weak or dead germs seemed to cause a slight case of the disease, but also provided protection if the animal contracted a stronger dose of the same disease later.

At first, Pasteur believed he could control all kinds of infectious diseases with injections of stale cholera germs. He soon discovered that cholera vaccinations protected only against cholera. Other diseases were caused by specialized germs and had to be treated with cultures made from those specific germs.

Louis Pasteur's work with neglected and weakened cultures resulted in the development of vaccines. Today millions of people around the world are vaccinated to protect against diseases that once would have caused many deaths.

DID YOU KNOW?

ALTHOUGH PASTEUR WAS THE FIRST PERSON TO CULTURE GERMS FOR VACCINATIONS, THE USE OF VACCINES ACTUALLY STARTED ALMOST A CENTURY EARLIER THANKS TO ANOTHER CHANCE INCIDENT. EDWARD JENNER, AN ENGLISH DOCTOR, OVERHEARD A MILKMAID TALKING ABOUT THE DEADLY DISEASE, SMALLPOX. "I CANNOT HAVE THAT," SHE SAID, "FOR I HAVE HAD COWPOX."

JENNER REALIZED THAT THERE WAS TRUTH TO THE MILKMAID'S WORDS. COWPOX WAS A MILD DISEASE SOMETIMES SPREAD FROM COWS TO THOSE WHO TOUCHED THEIR UDDERS. THOSE WHO CAUGHT COWPOX NEVER SEEMED TO CATCH MORE DEADLY SMALLPOX. IS SOMEONE INFECTED WITH COWPOX PROTECTED FROM SMALLPOX? JENNER WONDERED.

TO TEST HIS IDEA, JENNER NEEDED A HUMAN SUBJECT. PEOPLE OF HIS TIME SO FEARED SMALLPOX THAT THE PROMISE OF BEING IMMUNE FROM IT SEEMED ATTRACTIVE. JENNER PERSUADED THE PARENTS OF A YOUNG BOY TO ALLOW THEIR SON TO PARTICIPATE IN AN EXPERIMENT. FIRST JENNER SCRATCHED THE BOY'S SKIN WITH FLUID FROM THE SORES OF A COWPOX VICTIM. THREE WEEKS LATER JENNER SCRATCHED THE BOY'S SKIN AGAIN, THIS TIME WITH A SMALL AMOUNT OF SMALLPOX FLUID. AS EXPECTED, THE BOY DID NOT BECOME INFECTED.

TO EXPLAIN THE DISEASE-FIGHTING METHOD, JENNER COINED THE WORD "VACCINATION" FROM THE LATIN WORD VACCA, MEANING "COW."

John and Will Kellogg – 1894

A STICKY MESS

Millions of people start each day with a bowl of cereal flakes. John Harvey Kellogg would be pleased. So would Will Keith Kellogg. The two brothers invented the stuff.

In the late 1800s, Dr. John Harvey Kellogg operated a medical boarding house in Battle Creek, Michigan. During their stay, patients were expected to follow Dr. Kellogg's prescription for health — plenty of fresh air, exercise, a good night's rest, and a diet free of coffee, alcohol, spices, and meat. Most people had no trouble with the fresh air and exercise parts, but patients accustomed to spicy meat dishes often complained that the vegetarian meals tasted bland.

Always on the lookout for ways to make his food taste better, Kellogg and his younger brother, Will Keith, experimented with various grain mixtures. They spent evenings in the hospital kitchen boiling, mashing, and baking nuts and grains. Boiling removed starch and created new flavors and textures, but it also made grain gooey and gummy. No matter how many times they tried, the Kelloggs were left with a sticky mess that baked into disgusting, doughy globs.

One evening while they were boiling yet another batch of grain, the Kellogg brothers were called away on urgent business. They hurried out of the kitchen, leaving the pot to cool on the stove. By the time they returned to their experiment two days later, the over-boiled mush had started to dry and go moldy.

Instead of throwing it out, they continued their experiment. They passed the dried dough through rollers to flatten it. To their surprise, each grain formed a separate flake and each flake toasted evenly in the oven. With more experimenting, the Kelloggs found the perfect formula for boiling and waiting that produced light, tasty, and unmoldy flakes.

From a health point of view, Dr. Kellogg was satisfied with the flakes. Not so Will. He saw business opportunities in their discovery. Eventually, he bought his brother's share of the flake invention, processed and packaged the cereal in Battle Creek, and created the Kelloggs food empire we know today.

MORE ACCIDENTAL DISCOVERIES

In 1928, while tinkering with new gum recipes in his spare time, accountant Walter E. Diemer of the Fleer Chewing Gum Company in Philadelphia produced a batch with unusual properties. It was less sticky and more elastic than regular chewing gum, stretched easily, and could be blown into bubbles. Diemer looked for a way to tint the bland looking gum to make it more appealing. The only food coloring available in the company lab was a shocking pink. As a last resort, Diemer added it to the mix. Marketed by the Fleer Chewing Gum Company as Dubble Bubble and sold for a penny a piece, the pink bubble gum was wildly successful. "It was an accident," Diemer said later. "I was doing something else and ended up with something with bubbles."

Wilhelm Roentgen – 1895

GLOW IN THE DARK

On November 8, 1895, Wilhelm Roentgen hurried through an experiment so he could get home in time for dinner. He wrapped a sheet of heavy black paper around a long, empty glass cylinder known as a cathode ray tube. He prepared a chemically treated cardboard screen to place near it. Then he hooked up the terminals of a power source to two metal plates inside the cathode ray tube and turned off the lights in the room.

Cathode ray tubes produce a type of radiation known as cathode rays which cause the glass walls of the tube and the air around it to fluoresce or glow an eerie lime-green color. Certain chemicals held close to the tube fluoresce as well, but the range is short – a few centimeters at best. Any farther away from the tube and the green glow vanishes.

That November day, with time marching ever closer to the dinner hour, Roentgen hurried through the last steps of his experiment hoping to discover more about the rays' properties. Would cathode rays penetrate the heavy paper sheathing he'd wrapped around the tube? Would they cause the chemically treated cardboard screen to glow? Based on his previous experiments, Roentgen expected they would.

When Roentgen shut off the lights and turned on the power, no light eked through the paper shield. There was no halo of green around the tube and none from objects nearby. This surprised Roentgen. Then, almost immediately, he realized he'd made a mistake. In his hurry, he'd forgotten to set up the cardboard screen. No wonder.

Roentgen glanced back. Where did he leave the piece of cardboard? A ghostly green light shimmered from far across the dark room. Roentgen shut off the power to the cathode ray tube. The glow immediately disappeared. When he turned on the current again, the green glow returned.

Curious about the source, Roentgen lit a match. The glow was coming from the cardboard screen that he'd mistakenly left lying on a bench. Roentgen knew that cathode rays couldn't travel that far. The tube had to be giving off another type of radiation, one with different properties than cathode rays possessed.

Throughout the night and during the following day, Roentgen performed many experiments with the mysterious ray. He knew that it could travel through the air and through paper, but could it travel through other objects, too?

It did. Glass, wood, rubber, and other materials could not block the ray. It traveled through them as if they weren't even there. Only one metal seemed to stop it – lead.

But the most surprising property of the ray was discovered by chance. One day while he had the cathode ray tube switched on, Roentgen moved a small piece of lead into its path. On a screen behind the object, he saw its shadow. But there was something else, too – an outline of the bones in Roentgen's hand. The rays had penetrated human flesh.

Roentgen asked his wife to help in his next experiment. She held her hand between the cathode ray tube and an unexposed photographic plate. When Roentgen developed the plate, he saw a permanent picture of the bones of his wife's hand imprinted on it, surrounded by a dim outline of the flesh. Terrified that she might be seeing the ghost of her own hand, his wife refused to participate in any further experiments.

Because so much was unknown about the rays, Roentgen called them X-rays, a name we still use. Today Wilhelm Roentgen's mysterious ray has many uses, from detecting flaws in the welded joints of spaceships to enabling doctors to "see" into the human body to look for broken bones.

DID YOU KNOW?

ONE OF THE MOST COMMON MODERN X-RAY MACHINES IS THE SECURITY SCANNER FOUND AT AIRPORTS. BY USING LOW-LEVEL X-RAYS, THE SCANNER ILLUMINATES PURSES AND SUITCASES, SHOWING THEIR INSIDES WITHOUT DAMAGING THE CONTENTS.

Henri Becquerel – 1896
MYSTERIOUS IMAGES

For his experiment, Henri Becquerel, a French scientist, needed only a sunny day and a few supplies – a sheet of black paper . . . an unexposed photographic plate . . . a crystal of uranium salt.

Becquerel's experiment was pretty basic. He wrapped the photographic plate in the black paper and placed it in bright sunlight. On top of the paper, he set a crystal of uranium salt – a radioactive substance that fluoresced or glowed a strange blue color in sunlight. To see if radiation had penetrated the paper, Becquerel developed the photographic plate later by passing it through a series of chemical baths.

Although the experiment was simple in design, Becquerel hoped to answer complex questions. What caused uranium to fluoresce? What properties did its radiation possess?

Becquerel figured that sunlight triggered uranium's reaction, and his experiment was designed to prove it. Because the photographic plate was enclosed in black paper, sunlight could not penetrate it. Anything appearing on the plate must come from the glowing crystal that seemed to fluoresce only when sunlight was present.

After each sunny day, when Becquerel developed the plate, he found – as expected – a well-defined image of the crystal. He believed he was well on his way to proving the importance of sunlight to uranium's radiation.

On February 26, 1896, Becquerel repeated his experiment. But that day and the next two, the skies were so cloudy that the crystal hardly glowed at all. Impatiently, Becquerel developed the plate anyway expecting to find only a faint image of the crystal.

To his surprise, the crystal's image was as sharp as ever. Sunlight did not seem to have any effect on the amount of radiation or on the strength of the rays the crystal produced.

Quickly Becquerel set up another experiment. This time, he placed the wrapped photographic plate with the crystal on top inside a dark cupboard. When he developed the plate a few days later, a clear impression of the crystal showed again. Even without sunlight, without glowing at all, the crystal still emitted radiation.

Becquerel's accidental discovery proved that some materials are naturally radioactive. Unaided by the sun or other sources of energy, they emit invisible rays all on their own.

Several years passed before the importance of Becquerel's discovery was fully recognized, but his work opened up a new branch of science – the study of the energy locked inside radioactive substances such as uranium. It was the beginning of the nuclear age.

Alexander Fleming – 1928
CLEAR RINGS, SHRINKING PATCHES

In September 1928, having just returned from holidays, Dr. Alexander Fleming inspected his crowded laboratory. Set inside a London hospital, the lab was a clutter of tables and shelves holding bottles, beakers, and dozens of small, flat plates called culture dishes. Inside each culture dish, millions of bacteria grew in colorful patches on a jelly-like substance. The jelly was their food source, and encouraged by warm conditions in the laboratory, the bacteria thrived.

The cultures of bacteria were the heart and soul of Fleming's research, and having been away for a few weeks, he drilled his assistants with questions. How were the bacteria doing? Have there been any changes?

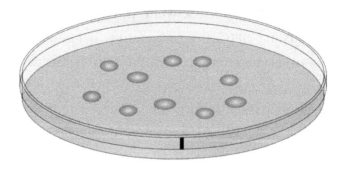

The news Fleming received disappointed him. One night while he was away, someone mistakenly left a window open. Bacteria from outside infiltrated the laboratory. Now the culture dishes were contaminated. There was no way of knowing just what kind of bacteria grew on the plates, nor what conditions aided or hampered their growth. The doctor's work was ruined. He'd have to start over.

The news hit Fleming hard. Reluctantly he tossed dishes into the garbage, a dreary, disheartening task, interrupted only when one assistant walked into the room. Fleming paused long enough to show the assistant a contaminated plate. Shame, isn't it? What a mess.

Suddenly Fleming stopped. He looked closer at the plate in his hand. "That's funny," he said. Mold much like the one sometimes found on bread or cheese grew along the top edge of the plate. In a ring around the mold, Fleming noticed a clear space. Bacteria grew right up to the ring, but not past it, and the patches of bacteria closest to the ring seemed to be shriveling and shrinking. Whatever the mold was, it appeared to have mysterious powers over bacteria.

With the discovery of this strange mold, Dr. Fleming began a new series of experiments. He took a speck of mold off the dish and put it in some broth where it multiplied and grew. Days later, he suctioned off a bit of the broth and added it to other culture plates. While some bacteria on the culture plates shriveled and died, other patches remained unaffected. The broth seemed to target certain kinds of bacteria, but not others.

Other experiments followed. Fleming identified the mold as penicillium. He compared it to other kinds of mold and tested different types in culture dishes containing bacteria. None of the other molds had the same effect as the penicillium mold.

Fleming also injected some of the penicillium broth into a rabbit. The rabbit remained healthy, an encouraging sign that suggested penicillium might be safe for human use, too.

After years of research, Fleming was able to extract a drug from the special mold. Because it came from the penicillium mold, he called it *penicillin*. Today penicillin is one of the most commonly used and effective disease-fighting antibiotics available, capable of controlling or counteracting a number of infections.

Sometime later, Fleming acknowledged the role of chance in his discovery. "I have been wonderfully lucky," he said. Luck certainly did play a part in the discovery of penicillin, but Dr. Fleming deserves credit, too. Without his keen observations and years of patient research, the opportunity that fate provided would have been missed or thrown away.

DID YOU KNOW?

IN 1945, ALEXANDER FLEMING SHARED THE NOBEL PRIZE FOR MEDICINE FOR HIS DISCOVERY OF PENICILLIN. IN AN ARTICLE HE WROTE, FLEMING GAVE THIS ADVICE: "NEVER NEGLECT AN EXTRAORDINARY APPEARANCE OR HAPPENING. IT MAY BE – USUALLY IS, IN FACT – A FALSE ALARM THAT LEADS TO NOTHING, BUT MAY, ON THE OTHER HAND, BE THE CLUE PROVIDED BY FATE TO LEAD YOU TO SOME IMPORTANT ADVANCE."

Julian Hill – 1934

STRETCHY STRANDS

In the 1930s, there was a worldwide shortage of natural fibers such as silk and cotton. Chemists at the DuPont Company set out to produce synthetic materials to take their place. Using coal, petroleum, and other ingredients, they tried to make artificial fibers that would be strong, but still soft and flexible enough to weave into cloth.

The DuPont chemists mixed batch after batch of new compounds. Some were too brittle and broke when stretched into threads. Others were too soft and sticky. Many of the failures were tossed out. Some were stored on shelves and forgotten.

One day, Julian Hill, one of the chemists, made an interesting discovery. He was stirring one of the too soft mixtures with a glass stirring stick. As he stirred, a small ball of material collected at the end of the stick. When Hill pulled the rod from the mixture, a long silky thread formed. The more he pulled, the longer the thread became.

Hill called other chemists over to have a look. One of them drew a sample from the container and pulled it. Sure enough, a thin thread formed. Excited talk filled the room as other scientists drew samples and tried the same thing. Then two of them played a game. They ran down the hall and stretched a ball of material into a thin strand the length of several rooms. The farther apart they went, the longer and stronger the fiber became.

Unexpectedly, Hill and his companions had discovered a new process called cold drawing. By pulling and stretching, they forced molecules in the fibers to realign, creating long chains of stronger material. The cold drawing process was the secret to creating synthetic fibers.

Du Pont scientists tried the same technique on some of the earlier samples that had been considered failures. Many of these could also be cold drawn to produce strong threads. One of them turned out to be particularly strong, yet strangely soft and delicate like silk. It was called nylon.

In history books, May 13, 1940, is sometimes called "Nylon Day." That day stockings made out of nylon instead of silk went on sale for the first time in New York City. Four million pairs were sold in the first few hours – a colossal success – but the event might never have happened had it not been for a lucky break and a little horseplay in a laboratory a few years before.

MORE ACCIDENTAL DISCOVERIES

In 1965, while attempting to develop a lightweight, heat-resistance fiber to reinforce radial tires, DuPont chemist Stephanie Klowek noticed something odd about one of her test tube samples. While the other solutions were clear and thick as molasses, this one was cloudy and thin as water. Later she said of the moment: "I thought, 'There is something different about this. This may be useful.'"

Tests showed Klowek was right. Fibers extracted from the solution proved to be unusually strong, stiff, and yet remarkably lightweight. In 1971, Du Pont released Kevlar, a new fabric woven from Klowek's unusual fibers. Used whenever super strength is needed, Kevlar can be found in a range of products from bulletproof vests and body armor to optic cables, military helmets, skis, and brake pads.

MORE EXPERIMENTAL TWISTS

Louis Daguerre (1835)
PHOTOGRAPHY

People of his time painted or sketched pictures, but Louis Daguerre hoped to capture images using chemicals. One day, after experimenting with a metal sheet covered in iodine, Daguerre carelessly left a silver spoon on its surface. Later when he removed the spoon, Daguerre discovered a faint image of it on the metal sheet. He was quick to realize that silver from the spoon had reacted with the iodine. He conducted a series of experiments with the chemicals and eventually found a way to "fix" images permanently on light sensitive surfaces. Daguerre's discoveries were the beginnings of a new type of picture-making – photography.

Alexander Graham Bell (1875)
TELEPHONE

Alexander Graham Bell, an inventor and also a teacher of the deaf, and his assistant, Thomas Watson, were experimenting with the telegraph when one of the keys in Watson's transmitter jammed. He plucked at the spring, trying to set it free. In another room far away with his ear pressed to the receiver, Bell heard a faint musical hum instead of the usual click-clack sound of the telegraph. The incident proved to Bell that sound could be duplicated and sent along a wire from one place to another. Bell abandoned the original experiment and spent his time creating a new invention – the world's first operational telephone.

Thomas L. Willson (1892)
ACETYLENE

In his search for an economical way to extract aluminum from compounds like aluminum chloride, Thomas L. Willson of Hamilton, Ontario, Canada, mixed lime and coal tar, popped the mixture into an arc furnace, and jacked the temperature up to 2800°C (5070°F).

When he opened the furnace door later, Willson discovered a new crystalline substance in the container. Wondering if it might be sodium, he tore off a small piece and dropped it into a bucket of water. Sodium reacts instantly and violently with water, but this substance foamed slowly and quietly, creating a froth of bubbles.

Willson's experiment was a two-for-one winner. The crystalline substance turned out to be calcium carbide, a substance that aided the aluminum extraction process and lowered its costs. The frothy bubbles – an accidental by-product of the water test – turned out to be acetylene, a highly flammable gas useful for all kinds of things from welding joints to producing fertilizers. Using Willson's process, both chemicals could now be produced simply, economically, and in large quantities.

CHAPTER 5

CLEVER CONNECTIONS

Masaru Ibuka, honorary chairman of the Sony Corporation, often roamed the company halls, examining new products as they were being produced. One day in 1978, he found technicians in one room working on lightweight, high-quality headphones. In another wing far down the hall, he found other technicians at work on another new Sony product – a small cassette tape player.

"Hmm," thought Ibuka, "Why not combine the two products? Why not use the headphones in place of the speakers in the cassette player, add batteries to provide power, and make a small, high-quality player that people could carry anywhere?"

Ibuka's creative thinking brought the world a brand new product. Called the Walkman, his portable player proved to be an instant best-seller and the forerunner to modern sound innovations like the handy iPod.

Science is peppered with examples of sudden connections like these. While they might seem accidental, there's usually more than just luck involved. Behind the scenes, there's a curious and creative mind at work.

Hans Lippershey – About 1600
CLOSE ENOUGH TO TOUCH

Many different stories have been told about the invention of the telescope. The one about Hans Lippershey has never been disproved and may well be true.

Lippershey was a maker of eyeglasses in Middelburg, Holland. One day he was standing in the doorway of his shop polishing some lenses he had just made. It was his custom to hold the finished lenses up to the light. That way he could see flaws in the glass more easily. Absent-mindedly, he held two lenses up to the light instead of just one and looked through both at the same time.

The church tower in the distance seemed to leap out at him. Startled, Lippershey almost dropped the lenses. After calming down, he held up the lenses and looked at the tower again. Sure enough, the tower appeared to be close enough to touch.

Lippershey noticed that he had used two *different* lenses. The one closest to his eye was concave or curved inward on one side. The lens closest to the tower was convex or curved outward on one side. By moving the lenses closer or farther apart, he could focus the image of the tower clearly.

At first, Hans Lippershey simply thought he had invented an interesting toy. He mounted the lenses on a board so his customers could view the church tower in the distance. Soon the popularity of his invention grew and his business boomed.

Eventually Lippershey refined his device. He enclosed the lenses in an adjustable hollow tube. He called it his *kijkglas* ("look glass").

On October 2, 1608, Lippershey applied for a patent for his "look glass" so that he would be the only person allowed to produce and sell his invention. He was refused. The government told him that his idea was not original enough for "too many people have knowledge of this invention".

It seems that other lens grinders in Holland had applied for patents on similar inventions at about the same time as Hans Lippershey.

Exactly who discovered this trick with the lenses – and how it was discovered – may always be a mystery, but Lippershey's accident might well be the forerunner to the telescopes, binoculars, and microscopes we use today.

DID YOU KNOW?

WHEN GALILEO GALILEI READ ABOUT LIPPERSHEY'S INVENTION, HE CREATED HIS OWN "LOOK GLASS". WITH THE CRUDE DEVICE, HE MADE A NUMBER OF DISCOVERIES: SUNSPOTS ON THE SURFACE OF THE SUN; THE FOUR LARGE MOONS OF JUPITER; THE RINGS OF SATURN; AND MORE THAN 100 NEW STARS IN THE MILKY WAY GALAXY. PEOPLE OF HIS TIME WERE INTERESTED IN MORE PRACTICAL USES FOR THE TELESCOPE, HOWEVER – LIKE SPOTTING INCOMING SHIPS FROM A POSITION MANY BLOCKS AWAY FROM THE HARBOR.

Luigi Galvani – 1786
TWITCHING LEGS

In a fairy tale, a frog might be a handsome prince in disguise. In science, a frog just might lead to an important discovery.

Around the year 1786, an Italian university professor, Luigi Galvani, was preparing to experiment with static electricity. Like other professors of his time, he used one of his rooms at home as his laboratory. His pupils gathered there for instruction.

One day Signora Galvani sat in the room to watch her husband. To pass the time before class started, she prepared a tasty meal of frog legs, a delicacy in Europe. Wielding a sharp steel knife, Signora Galvani skinned and sliced dead frogs and then laid the pieces on a zinc plate beside her.

When she was finished, Signora Galvani set the knife down on the plate. She watched the students as they entertained themselves by cranking a nearby electrostatic generator to shower the room with sparks.

From the corner of her eye, Signora Galvani noticed movement from the dish of legs. Astonished, she turned to watch them closely. Sure enough, the legs twitched as if they were still alive.

Intrigued, Signora Galvani continued watching and soon noticed a pattern to the movements. Only those parts of the legs that touched the knife blade resting on the edge of the metal dish twitched. The twitching also seemed to happen only when sparks were produced by the nearby electrostatic machine.

When her husband came home, she excitedly told him of her discovery. Fascinated by the twitching reaction, Galvani began a long series of experiments. He reasoned that if sparks from an electrostatic machine caused twitching then lightning should have to same effect. To find out, he hung frog legs from brass hooks on the iron railings that surrounded his house.

One sunny day, a light breeze pushed the legs against the iron railing. There wasn't a cloud in the sky or a hint of electricity, but the legs twitched anyway. Stumped, Galvani moved indoors to conduct new tests under more controlled conditions. He laid the frog legs on an iron plate and pressed the brass hook against it. Again, the legs flinched.

Galvani knew that muscles in a frog's leg twitched when they came in contact with electricity. Now the twitching happened even when electricity was nowhere nearby. Why? Galvani thought there must be a natural source of electricity inside the frogs. He called it "animal electricity".

Later another Italian professor, Alessandro Volta, also investigated the twitching. He found that when two different metals – the zinc plate and steel knife in Galvani's case – were separated by a moist conductor like frog's legs, electricity was produced.

Galvani may have been wrong when he believed that animals have a natural source of electricity, but his discoveries were valuable to others. They led to the invention of the battery, a device that produces power when two metals, usually copper and zinc, are separated by a moist mixture.

Ignaz Semmelweis – 1846

MYSTERIOUS DEATHS

In 1846, doctors at Vienna's General Hospital in Austria were faced with a puzzling problem. Why were so many mothers and babies in the maternity ward dying of childbed fever? And why was the death rate in one maternity ward many times higher than in another?

The hospital served many women who were charity cases. These women could not afford costly medical care. In return for medical attention for themselves and their babies, they agreed to be part of the training program for medical students. Surprisingly, the death rate in the training ward was ten times higher than in another ward where doctors rarely visited and babies were delivered by women known as midwives.

Determined to unravel the mystery of these strange deaths, Dr. Ignaz Semmelweis observed the wards and patients closely. With other doctors, he carefully examined the dead bodies in the hope of uncovering some clues.

One day there was an unfortunate accident. One of the doctors cut his finger as he dissected a dead body. Even though the cut was minor, the doctor soon felt ill. He developed a fever and in a few days died of blood poisoning.

Semmelweis noticed that the doctor's symptoms were suspiciously like those of patients who died of childbed fever. Acting on a hunch, he watched the movements of the doctors and students. An interesting pattern emerged.

Midwives who attended patients in the healthier ward where doctors rarely visited, did not examine bodies in the dissecting room. But doctors and students often went directly from the dissecting room to the other maternity ward with the higher death rate. None of them stopped to wash their hands before going from one room to the other.

All at once, pieces of the puzzle began to fit together. Semmelweis realized that doctors and students carried infection from the dead bodies into the maternity ward. He announced a new rule. From now on, patients, students, and doctors had to wash and disinfect their hands.

Just as Semmelweis suspected, the death rate soon dropped remarkably. Despite the success of his methods, other doctors ridiculed him. They refused to believe that such a simple procedure could solve the problem. Disgraced, Semmelweis was forced to leave Vienna. His rule was forgotten and again the death rate climbed.

Years later, doctors around the world admitted that Ignaz Semmelweis was right. Today hand washing is recognized as one of the necessary steps in preventing the spread of disease.

> **DID YOU KNOW?**
>
> AFTER HAVING WORKED FOR YEARS TO PREVENT THE SPREAD OF CHILDBED FEVER, IGNAZ SEMMELWEIS HIMSELF SUDDENLY CAUGHT IT AND DIED IN 1865.

Ivan Pavlov – 1901

MORE DROOL, PLEASE

At the Institute of Experimental Medicine in St. Petersburg, Russia, physiologist Ivan Pavlov was studying the digestive systems of dogs. In a series of carefully controlled experiments, he hoped to answer several questions. How did the amount of food eaten affect the quantity of saliva a dog produced? Did the type of food or the time it was served make a difference?

Normally, the sight or smell of food caused the dogs' salivary glands to swing into action, producing drool and beginning the digestive process. But one day Pavlov noticed something odd when an assistant entered the research area wearing a white lab coat. One of the dogs began to salivate even though the assistant wasn't carrying food. The same thing happened when an empty metal food cart was wheeled into the room – slobber, even when there was no food around.

Pavlov believed that the dog had been *conditioned* to salivate. Because assistants normally wore white coats during feedings and wheeled carts into the room, the dog had unconsciously learned to connect the sight of the coat and the clang of the cart to food. In time, just the white coat or the cart alone was enough to jumpstart the drooling process.

Pavlov set up a series of experiments where he paired food with various sights or sounds. In one, he rang a bell just before food was delivered. After a few tries, the sound of the bell itself was enough to cause the dog to drool.

Intrigued by the response, Pavlov devoted the rest of his life to studying conditioned learning. In 1904, the Noble Prize for Medicine or Physiology was awarded to Ivan Pavlov for his work on classical conditioning. His discovery laid the groundwork for a whole new branch of science – psychology, the study of human behavior.

MORE ACCIDENTAL DISCOVERIES

DOGS PLAYED A PART IN ANOTHER IMPORTANT DISCOVERY. IN 1889, GERMAN SCIENTISTS JOSEPH VON MERING AND OSCAR MINKOWSKI REMOVED THE PANCREAS FROM A DOG TO STUDY HOW THE DOG'S DIGESTIVE SYSTEM WORKED WITHOUT THE ORGAN. DAYS LATER, A LAB ASSISTANT NOTICED FLIES SWARMING AROUND A PUDDLE OF THE DOG'S URINE. CURIOUS, THE SCIENTISTS TESTED THE URINE AND FOUND IT CONTAINED HIGH CONCENTRATIONS OF SUGAR. IN TIME, SCIENTISTS DISCOVERED THE REASON. THE PANCREAS SECRETES A SUBSTANCE – INSULIN – THAT HELPS THE BODY METABOLIZE SUGAR. THE DISCOVERY WAS AN IMPORTANT STEP TO UNDERSTANDING DIABETES AND FINDING WAYS TO REGULATE THE DISEASE.

Clarence Birdseye – About 1917

FRESH FROZEN

Between 1914 and 1917, Clarence Birdseye lived as a fur trader in Labrador, Canada. Fresh food was not always available in the sub-zero climate, but Birdseye noticed that the native Inuit rarely went hungry. After a fishing or hunting expedition, they stored part of their catch outdoors. In the cold, dry air, food froze quickly. Months later, as the need arose, the Inuit thawed and cooked their frozen catch.

Probably hundreds of fur traders had observed this before, but Clarence Birdseye was the first one to take it a step further. He noticed that meat frozen on the coldest days tasted fresher and more tender than meat frozen on milder days. To find out why, he examined the meat under a microscope and compared the cells of samples frozen at different temperatures.

Birdseye noticed an interesting pattern. On milder days, when meat froze slowly, long thin ice needles developed that punctured the cell walls. Later, when the meat was thawed, the broken cell walls collapsed. Fluid seeped out, and the food tasted soggy and bland. On very cold days, though, meat froze so quickly that there wasn't time for needle-like crystals to form. When the meat thawed, the cell walls remained unbroken. The food tasted fresh and firm.

Birdseye ran tests on other types of food. He soaked a few cabbages in salt water and set them outside in the freezing wind. The cabbages froze quickly, their cells unbroken by ice crystals. When cooked, the vegetables retained their fresh-picked flavor.

Now that he understood how freezing preserved meat and vegetables, Birdseye tackled another problem. How could he provide people living in warmer climates with tasty fresh-frozen food?

When he returned to the United States in 1923, Birdseye experimented with rabbit meat and fresh fillets in his own kitchen. Later he worked in a refrigeration plant in New Jersey. Eventually, he invented a freezing machine. In it, very cold salt water was passed over metal plates that touched the cartons of food, freezing them in minutes instead of hours.

By the end of the 1920s, Birdseye began selling packages of his frozen foods. People soon discovered that the quick-frozen food tasted as fresh as the day it was frozen.

The frozen food business made Clarence Birdseye a wealthy man. His success must have given him the urge to invent as well, for by the time he died in 1956 at the age of seventy-three, he had over three hundred other inventions to his credit.

MORE ACCIDENTAL DISCOVERIES

LEGEND HAS IT THAT MAPLE SYRUP WAS DISCOVERED ACCIDENTALLY WHEN A WOMAN USED THE SAP FROM A MAPLE TREE IN PLACE OF WATER IN A MEAL SHE WAS PREPARING. AS THE FOOD COOKED, THE SAP BOILED AND THICKENED, ADDING A SWEET FLAVOR TO AN ORDINARY MEAL.

Arthur Fry – 1974

A REMARKABLE FAILURE

Spencer Silver figured his experiment was a failure. So did other scientists that worked for the Minnesota Mining and Manufacturing Company (3M) in 1970.

Silver tried to invent a new super-strong glue. Instead, the batch he mixed was the opposite – super-weak. The glue barely stuck, and it was so temporary that the two objects could be peeled apart easily.

The glue was labeled a failure, shelved, and almost forgotten.

Then one Sunday four years later, another 3M scientist, Arthur Fry, encountered a problem while singing in his church choir. The bits of paper he used to mark his place in the choir book often fell out. Fry kept losing his place, a frustrating experience.

"I don't know if it was a dull sermon or divine inspiration," Fry said later, "but my mind began to wander and suddenly I thought of an adhesive that had been discovered several years earlier . . ."

Fry remembered Spencer Silver's seemingly useless glue. Would that solve the problem? When he returned to work the next day, Fry tested his idea. He spread the super-weak glue on bits of paper and stuck them to the pages of his book. The markers stayed in place but separated with little effort.

For nearly a year and a half, Fry perfected the glue, adjusting the formula so that the markers peeled off without leaving a residue. When he was ready, Fry passed out samples to his co-workers at 3M. They weren't impressed. No one was sure why people would buy sticky note paper when ordinary scratch paper sold for so much less.

In 1977, Post-It Notes, as the sticky pads were called, were test-marketed in four cities. In two cities, sales were poor. In the other two, sales were amazing. When 3M representatives looked closer, they discovered the reason for the difference. In the two cities with terrific sales, dealers had passed out free samples. Once people had Post-Its in their hands, they discovered many different uses for the sticky paper.

Today Post-Its can be found in homes and offices, in many colors and designs, and on everything from refrigerators to television screens, proving that even failures can be outstanding successes given the right circumstances.

MORE CLEVER CONNECTIONS

Jean-Francois Champollion (1799)
THE ROSETTA STONE

An engraved stone found embedded in an ancient wall near the town of Rashid (Rosetta), Egypt, puzzled the French soldiers in Napoleon's army that found it. The stone was inscribed in three different scripts – hieroglyphic, demotic (native Egyptian), and Greek. Scholars who examined the stone realized that each script seemed to repeat the same message. Of the three scripts, though, only Greek was a living language. The ability to decipher hieroglyphics and demotic Egyptian had vanished a thousand years earlier.

Figuring that that the Greek script might serve as a decoder for hieroglyphics, scholars tried to match it to the symbols or 'glyphs' on the stone. Years of study finally led to a breakthrough in 1822 when French scholar Jean-Francois Champollion realized that glyphs represented sounds, not words as others had assumed. By cracking the Rosetta stone, Champollion gave historians a way of decoding thousands of years of lost history.

David Parkinson (1940)
GUN DIRECTOR

In the spring of 1940, engineer David Parkinson went to sleep upset by news of the day. It was wartime and things were not going well for the Allied forces. Thousands of soldiers had just been killed by German planes in a major battle off the coast of France.

During his sleep, Parkinson had a vivid and disturbing dream. He was in a gun pit with Dutch soldiers who were manning an anti-aircraft gun. Enemy planes screamed across the sky. The soldiers fired shot after shot at the attacking planes. With each shot, a plane plunged to the ground. Puzzled by the gun's accuracy, in the dream Parkinson moved in for a closer look. That's when he noticed a small round device mounted on the gun.

Parkinson awoke with the dream still clear. He realized that the round device was familiar to him. It was a potentiometer, an instrument he had used at work to control the movement of a machine called a strip recorder. Suddenly the dream took on new meaning. Perhaps the potentiometer could be used to control the movement of guns, too.

Parkinson's dream led to the development of the gun director, an invention that adjusted the aim of a gun by constantly recalculating the target's position. The gun director greatly improved the accuracy of anti-aircraft guns and helped the Allies turn defeat into victory.

Percy Spencer (1945)
MICROWAVE OVEN

When Percy L. Spencer, a self-taught British engineer, reached into his pocket for a chocolate bar after working on a radar set at the Raytheon Manufacturing Company, he discovered a gooey melted mess. Why? he wondered. The room hadn't been especially warm. On a hunch, he placed a few popcorn kernels near the magnetron, the radar set's power tube. In no time, the kernels popped. The next morning, Spencer brought a tea kettle to work. He cut a hole in the side of the kettle, placed a raw egg inside, and aimed the kettle at the magnetron. In seconds, the egg exploded, spewing bits of shell and yolk around the lab. To Spencer the conclusion was inescapable. The hidden source of energy that caused all of these had to be the magnetron. Spencer's observation led to the development of one of the kitchen's most popular items – the microwave oven.

Jocelyn Bell & Anthony Hewish (1967)
PULSARS

In July of 1967, Jocelyn Bell, an astronomy student at Cambridge University in England detected what she called 'a bit of scruff' while analyzing the data generated by a radio telescope. It was an unusually strange burst of energy that appeared and disappeared in a steadily repeating pulse every 1.3 seconds. The signal was unlike anything Bell or her advisor, Anthony Hewish, had noticed before. Was an intelligent life form trying to send a message? Half joking, they named the pulsating signal "Little Green Men".

Puzzled, Bell and Hewish investigated. Bell aimed the radio telescope at a completely different patch of sky. Again she detected a similar pulse. Then, over the Christmas holidays, she located two more, each from a different region of the sky, providing solid evidence that they had discovered a new brand of star – the pulsar.

Katherine Payne (1984)
ELEPHANT COMMUNICATION

While observing three Asian elephants and their calves at the Washington Park Zoo in Portland, Oregon, zoologist Katherine Payne noticed something unusual. The air around the elephant enclosure throbbed, sending a shudder through Payne's body. She recalled feeling the same sensation as a young girl when she stood near a pipe organ at church and the lowest notes on the organ were played. Were the elephants sending messages by making sounds too low in frequency for humans to detect?

Payne used electronic instruments to record the elephants. Her ground-breaking investigation proved that elephants communicate using hundreds of different calls, many of them at frequencies well below the range of human hearing.

CHAPTER 6

SURPRISE ENDINGS

Two hundred years ago John Spilsbury, a British teacher, tried a new learning aid in his classroom. Many of his students had difficulty remembering names and places on maps so Spilsbury invented a device to help them. He glued a map of England and Wales on to a thin piece of wood. Then he cut the wood along county boundaries. By reassembling the pieces, his students learned geography quickly.

Spilsbury thought his invention would be useful only in classrooms, but others saw possibilities he didn't. Colorful pictures were substituted for the maps. These were glued onto wafers of wood and then cut into odd-shaped pieces. People had fun fitting the pieces together to make up the complete picture. From John Spilsbury's educational invention came the jigsaw puzzle we still enjoy today.

As the stories in this chapter show, sometimes a person has one idea in mind, but then fate steps in to twist and change it. In the end a different product or plan emerges. Sometimes it's even better than the original.

Sixteenth Century

A POISONOUS DRINK

The Indian was lost, or so a story from the sixteenth century goes. Lost in the thick jungles of the high Andes in South America. Worse still he was sick with malaria, a deadly disease. Delirious with fever, he wandered for days. His head ached. His swollen tongue filled his parched mouth. Water was what he craved most – cool, refreshing water.

Miraculously, the man found it in a small stagnant pool hidden among the trees. He ran to the edge, threw himself on the ground, scooped up a handful of water, and gulped it down. Immediately he gagged and spit it out. The water was bitter tasting and clearly contaminated. A quick look around explained why. Cinchona trees grew at the pond's edge. Their roots reached into the water. The bark of the cinchona was poisonous, the man knew. Poison must have seeped into the water.

The man's thirst was so great that he didn't care. He choked down the foul-tasting water, drinking until he could drink no more. Then he waited for the poison to take hold. It never did. To the Indian's great surprise, his fever passed and he grew stronger – so strong that he was able to find his way back to his village. There he told his friends and relatives about the strange effects of the cinchona tree.

The news spread. When others who were sick with malaria drank potions made from the bark of the tree, they were cured, too. Eventually European missionaries heard about its unusual powers. They brought its bark back to their own countries and used it to cure thousands more.

The story of the Indian's discovery has been told and retold for hundreds of years. Many of its details fit facts that are well known. Centuries ago, malaria was a deadly killer with no known cure. Spread by mosquitoes, it struck young and old, killing more people that all wars combined. Then in the sixteenth century a miracle drug mysteriously surfaced from the forests of South America. *Quinine*, a chemical found in the bark of the cinchona tree – a tree which grew only in South America – was found to lower the fever and cure the disease.

Did a sickly Indian really stumble upon a cure by accident? Without written records and actual proof, we have no way of knowing for certain.

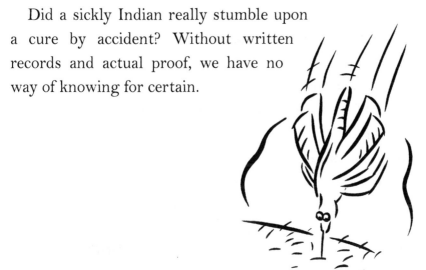

William Henry Perkin – 1856

STICKY MESS, VALUABLE SOLUTION

By all accounts, young William Henry Perkin was determined and clever. Hardworking, too. During Easter holidays in 1856, while others relaxed, the eighteen year-old chemistry student spent his days and evenings in a small, simple laboratory in his home in England. His goal was ambitious. From ordinary coal tar, Perkin wanted to produce something valuable and rare – quinine, the miracle drug that cured malaria.

Only one source of the drug existed. Quinine came from the cinchona tree which grew in far-off South America. Finding a cheap, practical way to produce quinine in a laboratory could save numerous lives and bring Perkin fame and fortune.

To Perkin, success was just a test tube away. He set up a series of experiments. He mixed coal tar with various chemicals, but none of the mixtures came close to having properties similar to quinine. Despite his disappointment, Perkin persisted. Each day he continued his experiments, varying chemicals, times, temperatures, and other factors in different attempts.

One day Perkin noticed a thick black residue at the bottom of a test tube. When he added alcohol to the sticky mess, it dissolved and a deep purple liquid appeared. Excited by the exotic color, Perkin abandoned his search for quinine. Instead, he tested the brilliant fluid.

At this time, the only dyes available were those that came from minerals, berries, flowers, and other natural sources. Perkin saw promise in the strange purple fluid. Here was a new hue that could be produced at will in a laboratory by combining common chemicals. Perkin named his lucky find *mauve*.

Perkin was fortunate in another way. By coincidence, a fashion trend for purple dresses had been started in Europe by Empress Eugenie of France. Perkin was just the man to supply large amounts of the new synthetic dye. He persuaded his father and brother to help him, and in a few months their factory was churning out great quantities of the artificial mauve color.

Was Perkin's success just a bit of luck? Not entirely. If William Henry Perkin hadn't been curious enough to stop for a second look at the black residue in the test tube, he would have overlooked his great discovery. As it was, his interest led to sweeping changes in the dye industry. Now colors could be obtained in greater variety, more cheaply and easily than ever before.

MORE ACCIDENTAL DISCOVERIES

In 1973, chemist Patsy Sherman accidentally spilled a chemical mixture on her shoe. The spot remained clean even when the rest of the shoe got dirty. The accident led to the development of Scotchgard, a stain-resistant mixture used to protect carpets, upholstery, and clothing.

Levi Strauss & Jacob Davis – 1873

PANTS NOT TENTS

In 1849, Levi Strauss, a young German-born salesman, packed up his wares, boarded a clipper ship in New York, and headed for California by way of Cape Horn in South America. Throughout the voyage, Strauss peddled his goods to fellow passengers. By the time the ship reached San Francisco, he had a pocketful of money and only a few rolls of canvas left unsold.

Strauss expected to sell the canvas for use as tents and wagon covers. But mining prospectors already had tents. Their pants had worn out, however, so it was pants that they wanted.

Strauss hired a tailor to fashion pants out of his leftover canvas. They sold in a flash. Almost overnight, other prospectors wanted Strauss's durable trousers so he opened a small manufacturing shop in San Francisco and began mass production. Later he switched from canvas to denim, a softer yet stronger material. To ensure that each piece of denim matched the others, he dyed them indigo blue. People started calling the pants "blue denims" and then eventually "blue jeans" (for Genoa, a city in Italy where a denim-like material was made).

The copper rivets that adorn Levi jeans came later after a brawny customer kept ripping the pockets and seams of his pants. Jacob Davis, a tailor in Reno, Nevada, reinforced the weakest points with rivets. The riveted pants were so popular that Davis partnered with Levis Strauss to apply for a patent to protect the product. The day they received the patent – May 20, 1873 – is considered by many to be the official birthday of blue jeans.

> ### DID YOU KNOW?
>
> A TRUCK DRIVER, MAKING A ROAD STOP IN 1998, DISCOVERED A BOX OF CLOTHES ABANDONED IN A RICKETY SHACK IN NEVADA. AMONG THE ITEMS INSIDE – A GRIMY, TATTERED PAIR OF JEANS FROM THE 1880S. AUCTIONED ON EBAY AS THE OLDEST LEVIS EVER, THEY WERE PURCHASED BY LEVI STRAUSS & COMPANY FOR A COOL $46,532.

Arthur Scott – 1907

TOO THICK, TOO HEAVY

In the early 1900s, the Scott Paper Company produced bathroom tissue and other paper products. To make tissue, the company ordered huge rolls of lightweight, absorbent paper from a paper mill. One day in 1907, someone at the mill goofed. Instead of the regular rolls, a large shipment of heavier-than-normal paper arrived at the Scott factory. It was too thick, too heavy, and too wrinkled to be used as bathroom tissue.

No one in the company knew what to do with the paper. Someone suggested sending it back. But Arthur Scott, head of the company, thought of something else. Why not perforate the thick tissue? Make it so that the paper tore off into towel-sized sheets then sell them as disposable hand towels.

Scott called the product Sani-Towels. At first, they were sold only to hotels, restaurants, and other places that had public rest rooms, but the towels proved so convenient and popular that in 1931 the company started manufacturing Scot Towels, a 200 sheet home version.

Today dozens of different brands of paper towels line supermarket shelves, each one ready to wipe, blot, dust, or polish just about anything.

MORE ACCIDENTAL DISCOVERIES

IN 1938, CHEMIST RAY PLUNKETT WAS EXPERIMENTING WITH NEW REFRIGERATOR COOLANTS WHEN HE UNEXPECTEDLY PRODUCED SOMETHING ELSE — A MYSTERIOUS WHITE POWDER. THE STUFF TURNED OUT TO BE THE SLIPPERIEST SUBSTANCE ON EARTH: TEFLON.

Lonnie Johnson – 1982

UNEXPECTED SPLASH

Lonnie Johnson was famous. Not just because he was an engineer who helped design three space probes for NASA. Not even because he held nearly forty patents on such practical inventions as thermostats and hair dryers.

No, much of Lonnie Johnson's fame came from a very different gadget. Carried by children and child-like adults alike, his best-known invention can hit a moving target at fifty paces with great accuracy, spraying refreshing relief on a hot summer's day.

In 1982, Johnson was developing an environmentally friendly heat pump for refrigerators, one that circulated water through tubes instead of relying on harsh chemicals. Armed with bits of plastic tubing and other spare parts, Johnson stationed himself in his bathroom to test his idea. He attached tubing to the faucet on the bathroom sink, rigged a homemade nozzle to the end, and turned on the tap. A stream of water shot across the room, hitting the shower curtains around the bathtub, blasting them back with surprising force.

The water traveled farther and faster than Johnson had imagined. "I knew it would make a neat water gun," he said later.

In his workshop, Johnson built a model. Using plastic pipes, Plexiglas, and an empty plastic beverage bottle as a storage tank for water, he constructed a rifle-like water shooter. Then he asked his six year-old daughter to test it out on their neighbors. The girl was happy to try it – what kid wouldn't be? – and the neighbors got totally soaked. Everyone agreed. The water gun was great fun.

It took almost four years for Johnson to receive a patent for the invention, and it wasn't until 1990 that the first Super Soaker hit the market place. Well over a billion dollars worth of Super Soakers have been sold since, making Johnson's bathroom surprise one of the best-selling toys of all time.

Michael Zasloff – 1986

THE FULLY RECOVERED FROG

For his research on genetic diseases, Michael Zasloff, an American scientist, operated on frogs to remove tissue samples that he would examine later. Usually, the operation was minor. Zasloff stitched up the frogs, returned them to the tank in a corner of his laboratory, and most of the time, the frogs recovered.

In the summer of 1986, Zasloff removed tissue from an African clawed frog. A few days later, he glanced into the murky waters of the tank to check on the frog's condition. The waters teemed with bacteria, a condition that could lead to infection. Zasloff half expected the frog to be dead, or at the very least sick. Instead, it was surprisingly active.

At first, Zasloff wondered if he had the wrong frog. He looked for the surgical wound. There it was in the frog's side, but instead of a red festering sore, the wound was small and almost healed.

Zasloff realized that he was observing a miracle. The frog had made a surprising recovery. Something must be protecting it from infection. But what?

Zasloff abandoned his earlier research and turned his attention to the frog. Eventually, he discovered the reason for the frog's unexpected good health. Special infection-fighting chemicals known as *magainins* were produced in the frog's skin glands.

Research is on-going, but we now know that the African clawed frog isn't the only frog species to produce magainins. Each species of frog produces a slightly different form, and with 5400 different frog species in the world, a whole army of disease-fighting chemicals might be lurking in ponds and lakes around the globe.

MORE SURPRISE ENDINGS

George Crum (1853)
POTATO CHIPS

George Crum had no intention of inventing a tasty new food snack. Revenge was what he really wanted. Crum was the chef at the Moon Lake Lodge in New York. One day a dissatisfied customer complained several times that Crum's french fries were not as thin, salty, or crisp as they should be. Unhappy about the complaints, Crum tried to get even. He made the potato slices so thin, salty, and crisp that he was sure the customer would hate them. To his surprise, the customer loved the dish. In his pursuit for revenge, George Crum created one of our favorite foods – the potato chip.

Joshua L. Cowan & Conrad Hubert (1900)
FLASHLIGHT

Joshua Cowen figured he had a whopper of an idea. Slip batteries into a slim metal tube, attach a light bulb to one end, stick the tube into a flowerpot, and place the pot on restaurant tables to create an instant centerpiece. The illuminated flower pot centerpiece never really caught on, and when Cowen ran into financial difficulties, he signed the rights over to businessman Conrad Hubert. Hubert saw new possibilities for Cowen's light-stick invention. He repurposed the light-stick, started the American Eveready Company, and sold millions of hand-held 'flashlights' and the batteries that powered them.

Charles Menches (1904)
ICE CREAM CONE

From his booth at the 1904 St. Louis World's Fair, Charles Menches sold assorted flavors of ice cream in dishes. Next to him, Ernest Hamwi sold a tasty Middle Eastern waffle-like pastry called zalabia.

August was hot, and Menches did a booming business selling ice cream. One scorching day, he sold so much ice cream that by noon he had run out of dishes. Without more dishes, he would be forced to close his booth. He glanced at his neighbor. Hamwi still had lots of zalabia left.

With Hamwi's help, Menches rolled the thin pastry into a cone shape, scooped ice cream on top, and passed it to his customers. They loved the combination of cool and crisp sensations, and Menches' ice cream cone was the hit of the exposition.

In 1912, Frederick A. Bruckman, an inventor from Portland, Oregon, created a machine that made pastry and folded it into cones. By 1920, one-third of all ice cream being consumed in the United States was eaten from cones.

James Wright (1945)

SILLY PUTTY

James Wright, an engineer working for General Electric, set out to find a way of making synthetic rubber by combining different chemicals. One of his mixtures – boric acid with silicone oil – ended up being sticky goop with strange properties. The stuff was soft enough to roll and shape, but also springy enough to bounce and stretch. The new compound was a poor substitute for rubber, but great fun to twist, mold, and toss. Eventually, it was sold in plastic egg-like containers as a children's toy. The playful substance – Silly Putty – was unbelievably popular when it first hit stores, and it continues to sell well today.

Norman Larson (1952)

WD-40

When an airplane manufacturer discovered that its planes were beginning to rust, it called upon Rocket Chemical, a small lubricant company, for help. On his fortieth try, head chemist Norman Larson found a formula that seemed to work. When sprayed on the planes, the product repelled water, coated metal, and protected parts from rust. Larson called it WD-40 (the WD standing for "water displacement").

Workmen at the airport discovered the stuff could be used on more than just airplanes. They claimed it fixed squeaks, loosened sticky parts, removed crayon smudges, dislodged gum from carpets, and did a host of other things. Rocket Chemical received so many orders for the wonder product that the company started selling WD-40 to the public. Today the WD-40 Company – the new name for Rocket Chemical – sells millions of dollars of the lubricant each year, and keeps an ever-expanding list of more than 2000 uses for its product on the 'Cool Stuff' section of its www.wd40.com website.

MORE ACCIDENTAL DISCOVERIES

THE ICE CREAM SUNDAE WAS INVENTED IN 1890 WHEN A WISCONSIN MERCHANT FACED A SHORTAGE OF ICE CREAM. TO STRETCH HIS SUPPLY, HE REDUCED THE ICE CREAM IN EACH SERVING. TO MAKE UP THE DIFFERENCE, HE ADDED CHOCOLATE SAUCE AND TOPPINGS.

GLOSSARY

accelerate – to move faster; to speed up

antibiotics – drugs used to kill harmful bacteria and cure infection

archives – a place which holds public records or historical documents

bacteriology – the scientific study of bacteria

behavioral psychologist – a person who studies the causes and effects of human behavior

cathode rays – high speed electrons emitted from a heated cathode ray tube

celluloid – a tough, flammable type of plastic

collodion – a syrupy, highly flammable solution used as an adhesive to close small wounds, hold surgical dressings, and for making photographic plates

concave – curved inwards; having a shape like the inside of a bowl

conductor – a material or device that transmits heat, electricity, or sound

contagious – infection or disease capable of being spread by contact between people

control group – in a test or experiment, a group that does not receive change and is being compared to a group that does receive change

convex – curved outwards; having a shape like the outside of a bowl

culture dish – a flat transparent dish used chiefly for growing microorganisms.

density – a measure of the compactness of a substance; the denser an object the greater its mass per unit volume.

electrostatic generator – a device that produces high voltage with a build-up of static electricity

extraction – the action of taking out something. Example: the dentist extracted a tooth

grafting – to join a shoot or bud with a growing plant by inserting it into the stem or another part of the plant.

immune – resistant to a particular infection or disease

implant – a device or material surgically placed in the body to replace or repair a malfunctioning part. Example: an implantable pacemaker

Leyden jar – a device that stores static electricity between two electrodes on the inside of a glass jar

magainins – chemicals acquired from the skins of frogs that have disease-fighting properties

midwife – a person trained to assist women in childbirth

nicotine – a toxic, oily liquid found in tobacco

nitroglycerine – a thick, pale yellow, unstable liquid that is explosive when jarred or heated suddenly

pacemaker – an artificial device for stimulating the heart muscle and regulating its contractions

paleoanthropology – the study of the origins the human species

pendulum – a weight suspended from a fixed support so that it can swing freely back and forth

potentiometer – a device that is used to control and adjust voltages in radios and TV sets

prototype – an early test sample or model of a future product

quinine – a bitter substance derived from the bark of the cinchona tree and used to treat malaria

radiation – energy transmitted in waves or a stream of particles. Light, heat, and sound are types of radiation

refractometer – an instrument used to measure the degree of bending a ray of light experiences as it passes from one medium such as air and into another medium such as glass

scientific method – a way of problem-solving that involves gathering data under carefully controlled conditions in order to analyze it and formulate conclusions

stethoscope – a medical instrument for listening to someone's working heart or lungs

transistor – a tiny electronic device that is used to control the flow of electricity in radios, computers, etc

vaccine – a preparation of weakened or killed bacteria that, when introduced into the body, makes a person less likely to catch a disease

vulcanized rubber – rubber that has been treated with sulfur and heat to give it greater durability, strength, and flexibility

FOR FURTHER READING

D'Estaing, Valerie-Anne Giscard. *The World Almanac Book of Inventions.* World Almanac Publications, 1985.

Flatow, Ira. *They All Laughed . . . From Light Bulbs to Lasers: The Fascinating Stories Behind the Great Inventions That Have Changed Our Lives.* New York: Harper Perennial, 1993

Goldsmith, Mike. *Eureka!: The Most Amazing Scientific Discoveries Of All Time.* Thames & Hudson, 2014.

Krois, Birgit. *Accidental Inventions: The Chance Discoveries That Changed Our Lives.* Insight Editions, 2012

Orzel, Chad. *Eureka: Discovering Your Inner Scientist.* Basic Books, 2014.

Roberts, Royston. *Serendipity: Accidental Discoveries in Science.* John Wiley & Sons, Inc., 1989

TIME 100 New Scientific Discoveries: Fascinating, Unbelievable, and Mind Expanding Stories. Time, 2011.

The following brand names, registered trademarks or patented names have been used in this book:

Aspartame
Birdseye
Dubble Bubble
DuPont
Harbitol
iPod
Ivory Soap Kellogg
Kevlar
Kodak
Levis
Lifesavers
Milk Duds
Plexiglas
Popsicle
Post-It Note
Sani-Towels
Scotchgard
Scot Towels
Silly Putty
Slinky
Super Soaker
Teflon
Velcro
Walkman
WD-40

INDEX

A

acceleration 11
acetylene 98
aluminum 97
Anning, Mary 56
Archimedes 6, 7, 8
Aspartame 13

B

bacteriology 13
battery 107
Beaumont, William 57, 59
Becquerel, Henri 88, 89
Bell, Alexander Graham 97
Bell, Jocelyn 119
Benedictus, Edouard 40, 41
benzene 22
Birdseye, Clarence 113, 114
blasting gelatin 39
blue jeans 131
brain function 64, 65, 66

C

cathode ray tube 85
cell phone 25
cell phone camera 72
celluloid 41
cereal flakes 82
Champollion, Jean-Francois 117
Charon 43
childbed fever 108, 110
cholera 78, 80
Christy, James 42, 43
cinchona tree 126
classical conditioning 112
cold drawing 94
collodion 39, 41
Cooley, Samuel 36
Cooper, Martin 24
Coover, Harry xi
COSTAR 20, 21
Cowen, Joshua 139
Crane, Clarence 16
Crocker, James 19, 20

D

Daguerre, Louis 96
de Mestral, George 23
density 8
diabetes 112
Diemer, Walter 84
digestion, process of 58
Dubble Bubble 84
dye, mauve 129
dynamite 38, 39

E

Eastman, George 70
Edison, Thomas 48
electrostatic generator 30, 32, 76, 105
elephant ommunication 120
Endo, Ichiro 50
Epperson, Frank 14
Etscorn, Frank 44, 45

F

Fife, David 62, 63
flashlight 140
Fleming, Alexander 90, 92
Franklin, Benjamin 47
frozen foods 113, 114

Fry, Arthur 115

G

Gage, Phineas 64, 65, 66
Galilei, Galileo 9, 11, 104
Galileo. *See* Galilei, Galileo
Galvani, Luigi 105, 106
Goldman, Sylvan 17
Goodyear, Charles 33, 35
grafting 61
Greatbatch, Wilson 49
Grimes, William 71
gun director 118
gunpowder 46

H

hair pillows 25
Hamwi, Ernest 140
Hargreaves, James 48
Hewish, Anthony 120
Hill, Jillian 93, 94
Howe, Elias Jr. 69
Hubble Space Telescope 19, 21
Hubert, Conrad 140
hygiene, hospital 109

I

Ibuka, Masaru 101
ice cream
  bar 70
  cone 140
  sundae 142
Ichthyosaur 56
inkjet printer 50
insulin 112

J

James, Richard and Betty 23
Jansky, Karl 43
Jenner, Edward 81
jigsaw puzzle 125
Johanson, Donald 67, 68
Johnson, Lonnie 135, 136

K

Kahn, Philippe 72
Kekulé, Friedrich 22
Kellogg, John and Will 82
Kevlar 95
Klowek, Stephanie 95
Koch, Robert 12, 13
Kodak xi, 70

L

Laennec, Rene 55, 56
Larson, Norman 141
laughing gas 36
Leyden jar 31, 76
Lifesaver 16
lightning rod 47
Lippershey, Hans 102, 104
Lucy 68

M

magainins 137
magnetron 119
maple syrup 114
McCrory, Phil 25
McIntosh, John and Allan 60, 61
McIntosh Red 60, 61
Menches, Charles 140
microwave oven 119
midwives 108
Milk Duds 16
Minkowski, Oscar 112
Morse, Samuel 53, 54
musical ratios 5

N

Neanderthals 68
Nelson, Christian 70
nicotine patch 45
nitroglycerine 38, 39
nitrous oxide 36, 37
Nobel, Alfred 38, 39
Nobel Prize 39, 92
nutrition 59
nylon 94

P

pacemaker 49
Parkinson, David 118
Pasteur, Louis 78, 81
Pavlov, Ivan 111, 112
Payne, Katherine 120
pendulum, principle of 9, 10, 11
penicillin 92
Perkin, William Henry 128, 129
photography 70, 96
Plunkett, Ray 134
Pluto 42, 43
Popsicle 14, 15
Post-It Notes 116
potentiometer 118
pulsar 120

Pythagoras 4, 5

Q

quinine 127, 128

R

radiation 85, 86, 88, 89
Red Fife 63
Roentgen, Wilhelm 85, 87
Rosetta Stone 117
rubber, vulcanized 34

S

Saccharin 13
safety glass 41
Sani-Towels 133
Schlatter, James 13
Schwartz, Berthold 46
scientific method 11
Scotchgard 130
Scott, Arthur 133
security scanner 87
Semmelweis, Ignaz 108, 109, 110
serendipity xii
sewing machine 69
Sherman, Patsy 130

shopping cart 17
Silly Putty 141
Silver, Spencer 115
Slinky 23
sound-recording 49
Spencer, Percy L. 119
Spilsbury, John 125
spinning jenny 48
stethoscope 56
St. Martin, Alexis 57, 58, 59
Strauss, Levi 131
sunscreen 26
Super Soaker 136

T

Teflon 134
telegraph 54
telephone 97
Temple of Mithras 71
Thomson, Elihu 76, 77

V

vaccines 80, 81
van Muschenbroeck, Pieter 30, 31
Velcro 24
von Mering, Joseph 112

W

Walkman 101
WD-40 142
welding, electric 77
Wells, Horace 36, 37
Willson, Thomas L. 97
Wright, James 141

X

X-rays 87

www.ingramcontent.com/pod-product-compliance
Lightning Source LLC
Jackson TN
JSHW011656201224
75788JS00007B/79